Preschool Math Workbook

Number Tracing, Addition and Subtraction math workbook for toddlers ages 2-4 and pre k

This book belongs to

_____

This is a beginner math workbook to help kids learn to write numbers, count, follow directions, learn patterns, practice addition & subtraction in a fun and engaging way.

This book is organized in a progressively skill building way for kids to develop confidence in math.

This book requires guidance from a teacher, parent or care giver to help the child follow the worksheets.

Meet Jojo.
Jojo is a curious elephant.
He loves to learn and play.
Learn to do math along with Jojo!

This Preschool Math workbook is divided into the following parts:

Part 1: Learning Numbers:
   Practice tracing numbers 1- 10
   Count and color objects

Part 2: Learning to Count
   Show the number with your fingers and write the answer
   Count and color objects

Part 3: Matching objects
   Before, Between & After Numbers
   More & Less

Part 4: Directions & Manipulatives
   Left & Right Differentiation
   Follow Directions
   Learn Patterns

Part 5: Learning Simple Addition
   Adding two numbers

Part 6: Learning Simple Subtraction:
   Subtracting two numbers

Kids can use a pencil, light color marker or highlighter to trace the dotted numbers. Use color pencils or crayons to color the objects.

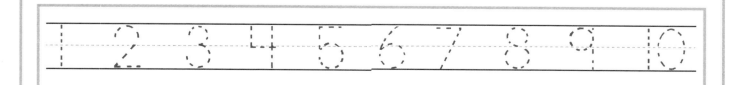

Hi!

My name is Sujatha Lalgudi. I sincerely hope you find my preschool math book to be helpful and fun.

Write to me at **sujatha.lalgudi@gmail.com** with the subject: **Preschool Math** along with **your kid's name** to receive:

- Additional practice worksheets.
- A name tracing worksheet so your kid can practice writing their own name.
- An Award Certificate in Color to reward your child!

If you liked this book, please leave me a review on Amazon! Your kind reviews and comments will encourage me to make more books like this.

Thank you
Sujatha Lalgudi

Part 1:

Learning Numbers

Count and color the objects.
Trace the numbers on the guided practice sheets.

Color 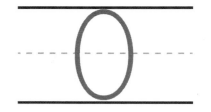 Zebra with O Stripes

zero

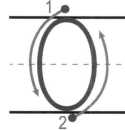 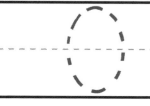 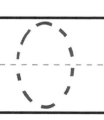 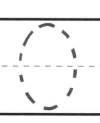

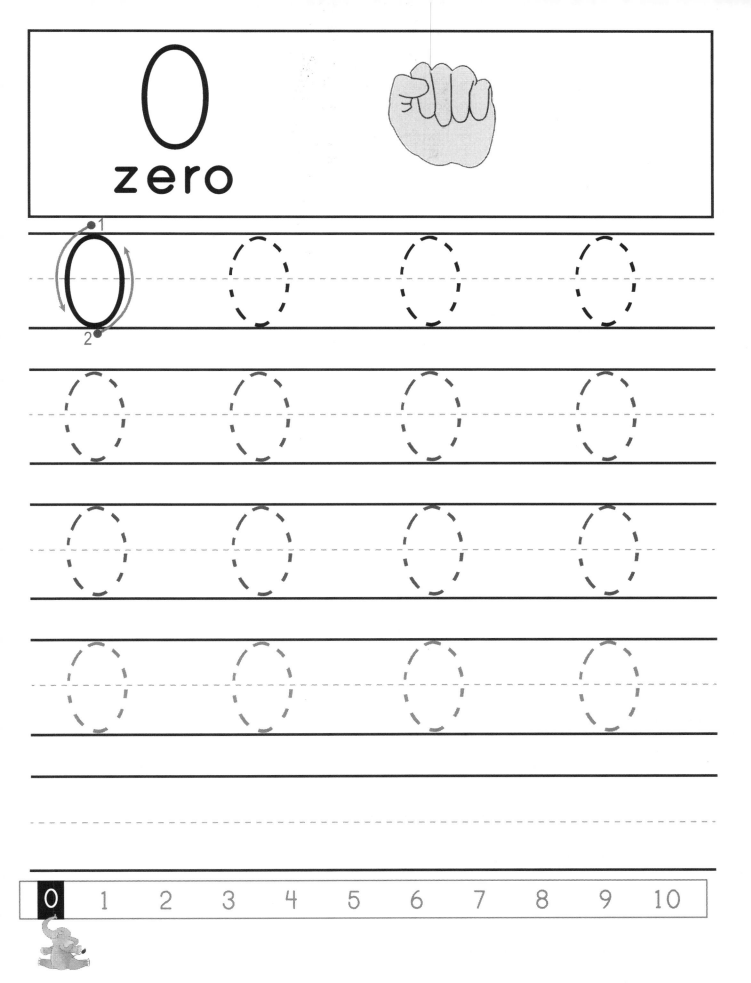

0
zero

0 1 2 3 4 5 6 7 8 9 10

Count
&
Color

I

Unicorn

one

|
one

1

0 1 2 3 4 5 6 7 8 9 10

Count
&
Color

2

Ants

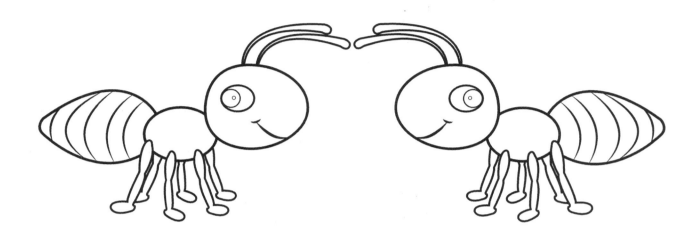

two

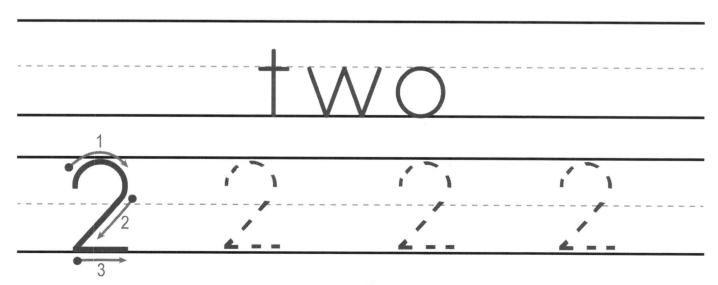

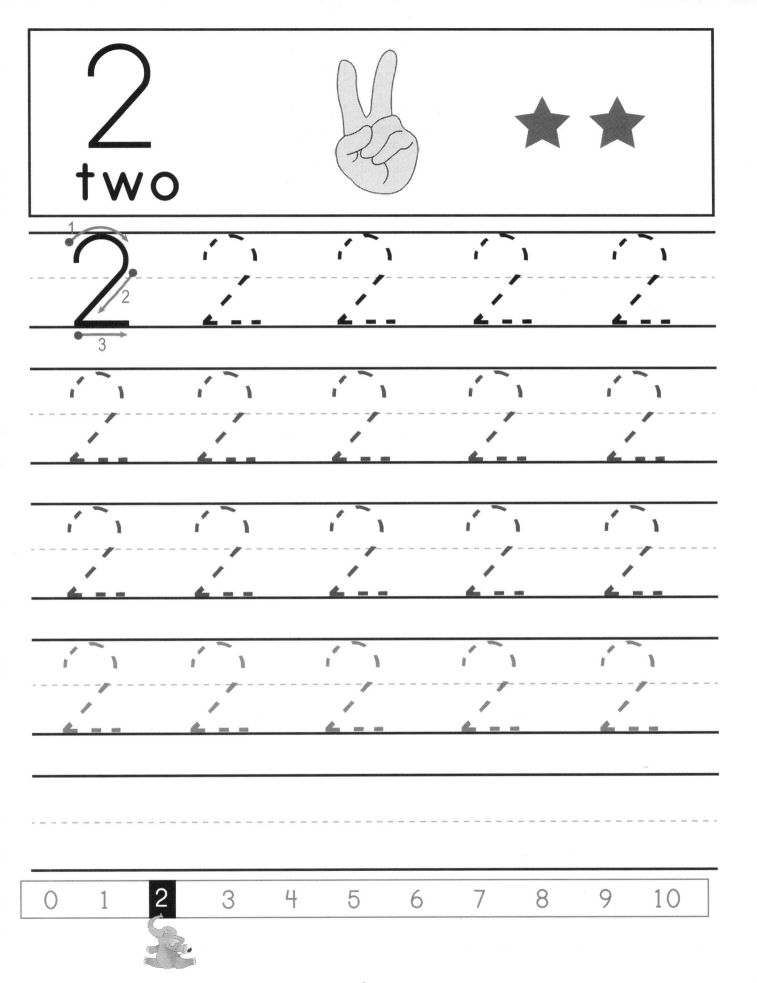

Count
&
Color

3

Frogs

three

3

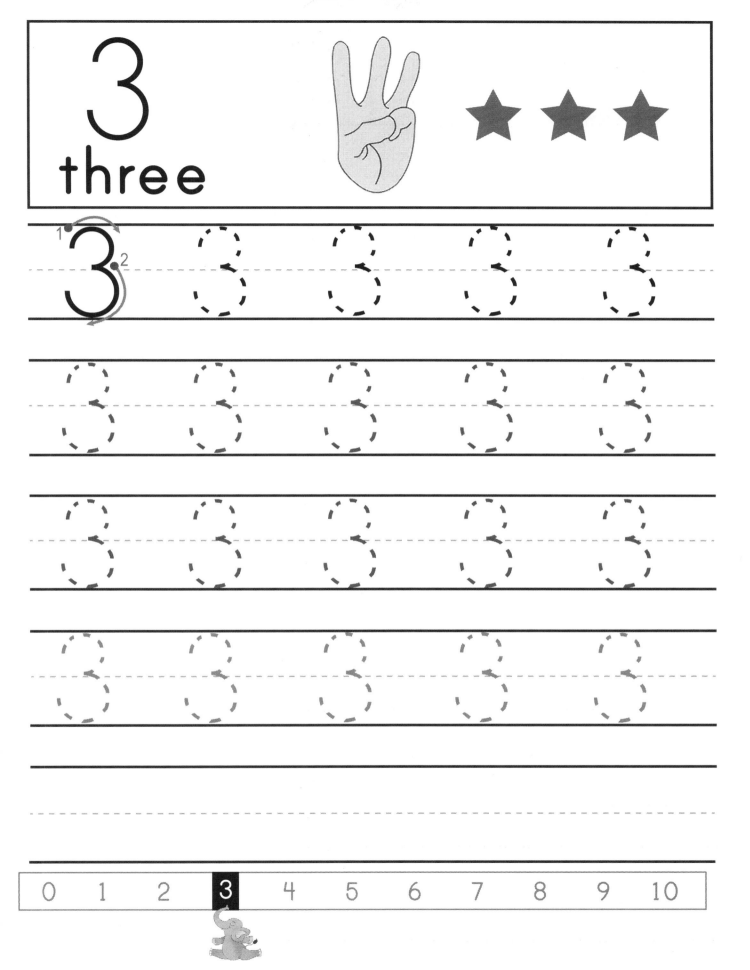

3
three

0　1　2　**3**　4　5　6　7　8　9　10

Count
&
Color

4

Butterflies

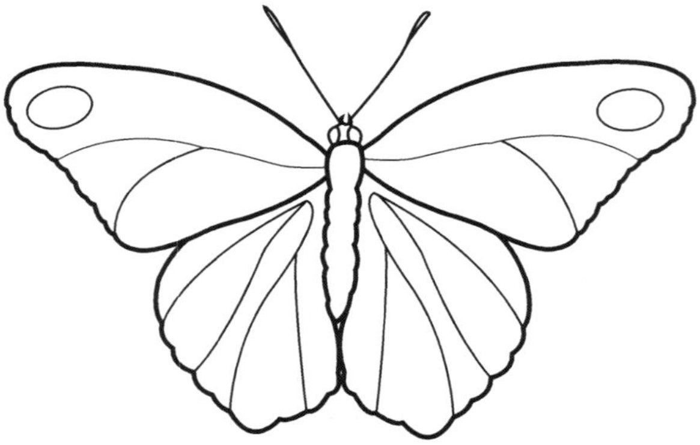

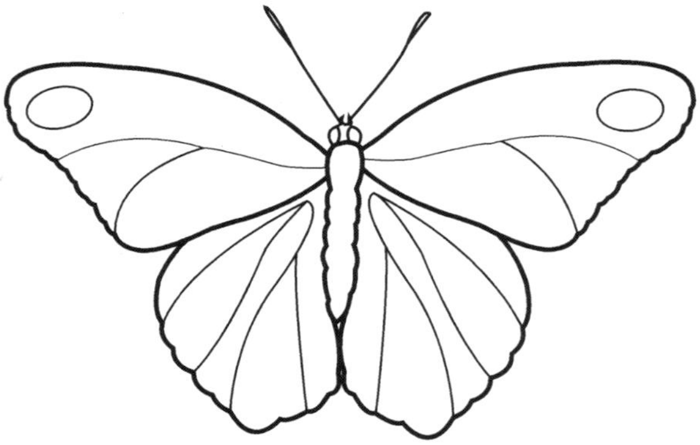

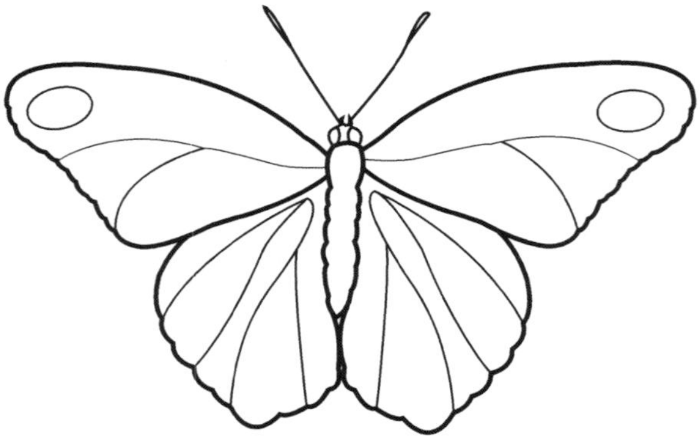

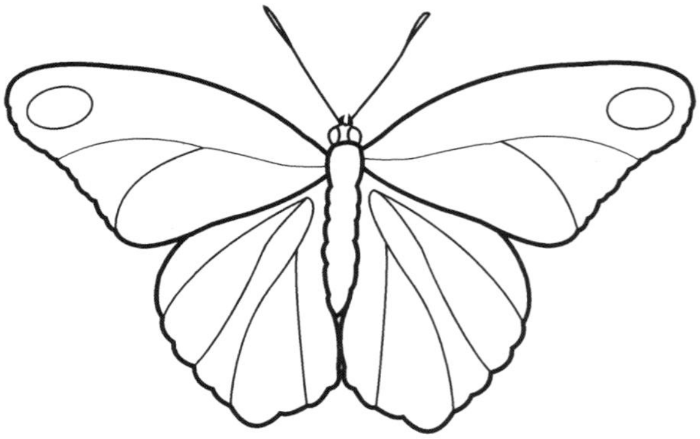

four

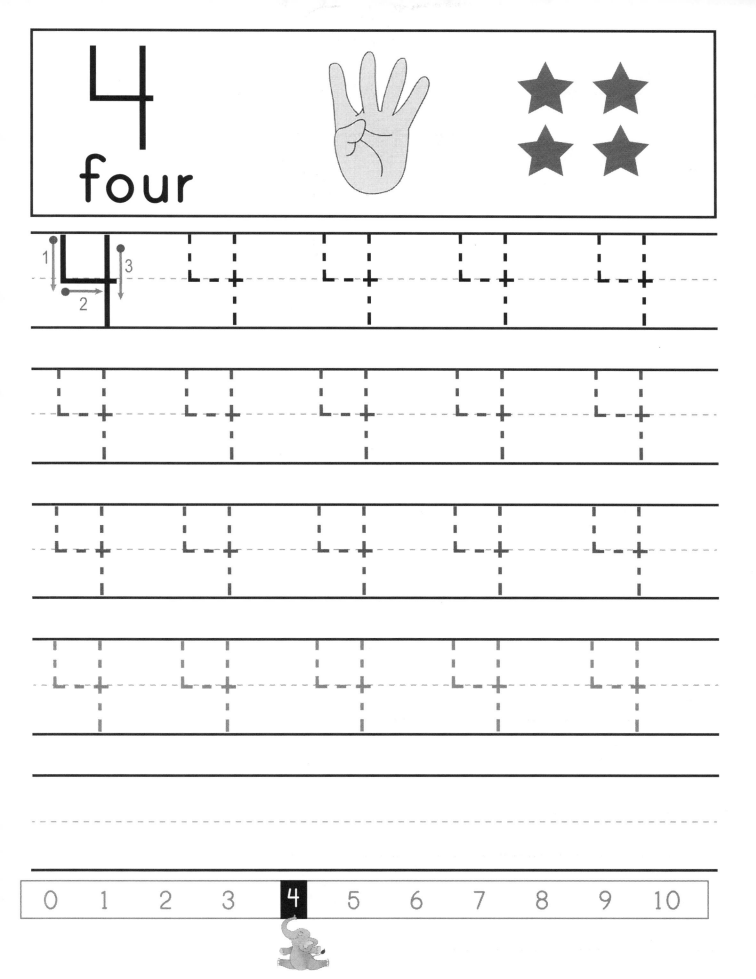

4
four

Count
&
Color

Orcas

five

5

5
five

0 1 2 3 4 **5** 6 7 8 9 10

Count
&
Color

6

Vicunas

six

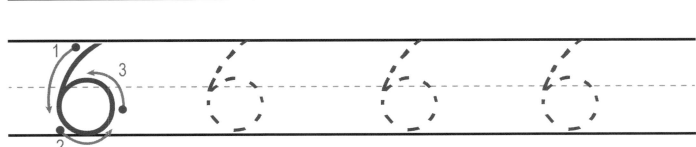

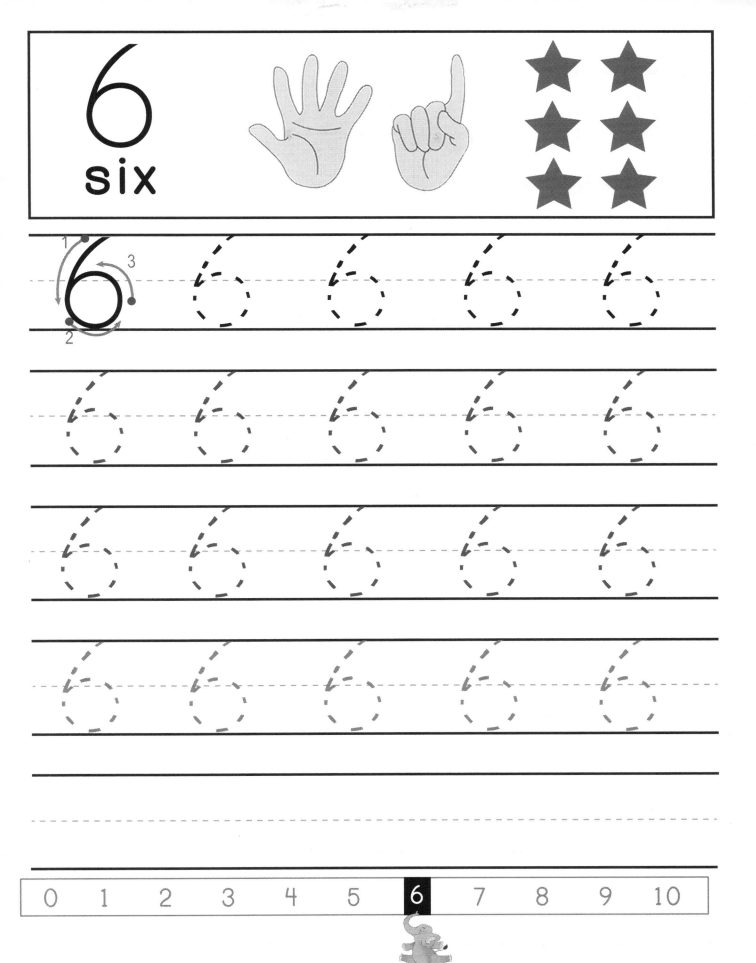

Count & Color 7 Crabs

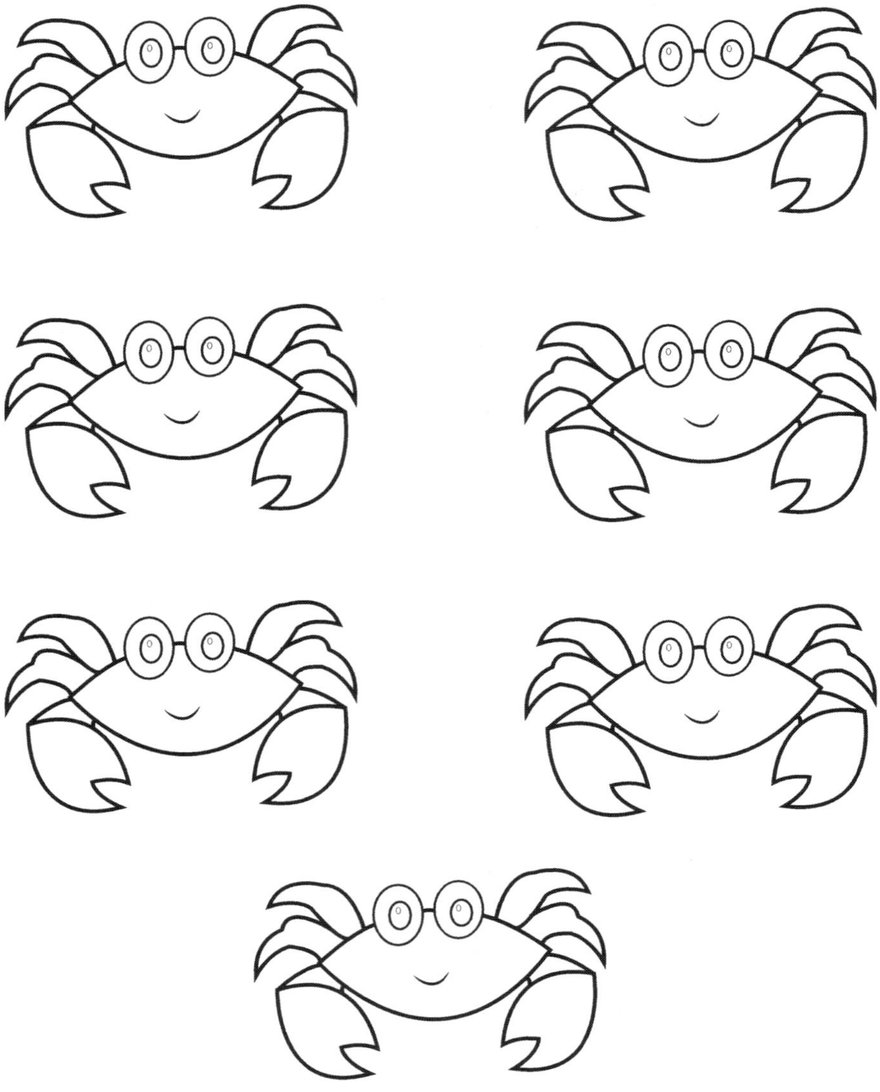

seven

1 2
7

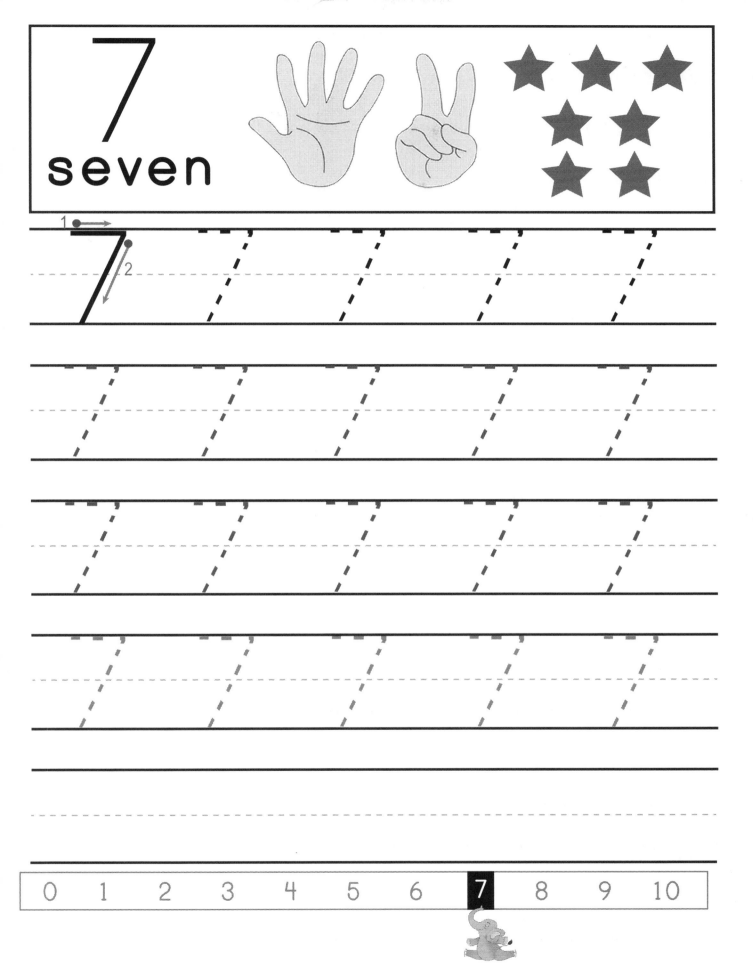

7
seven

0 1 2 3 4 5 6 **7** 8 9 10

Count
&
Color

8

Snails

eight

8

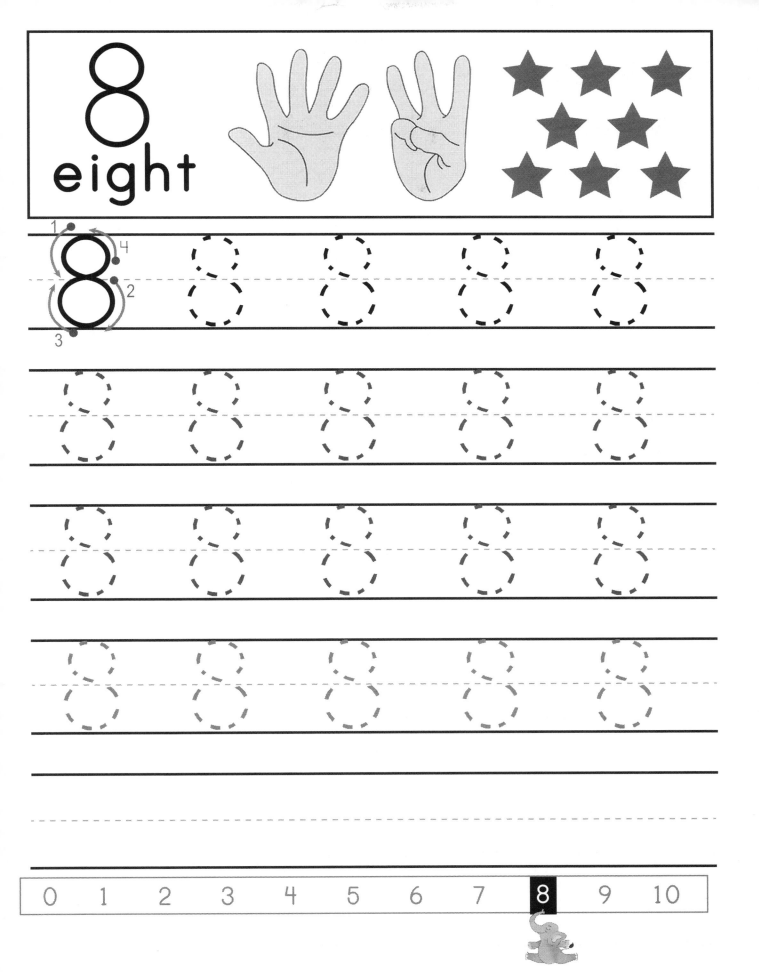

8
eight

0 1 2 3 4 5 6 7 8 9 10

Count
&
Color

q

Gifts

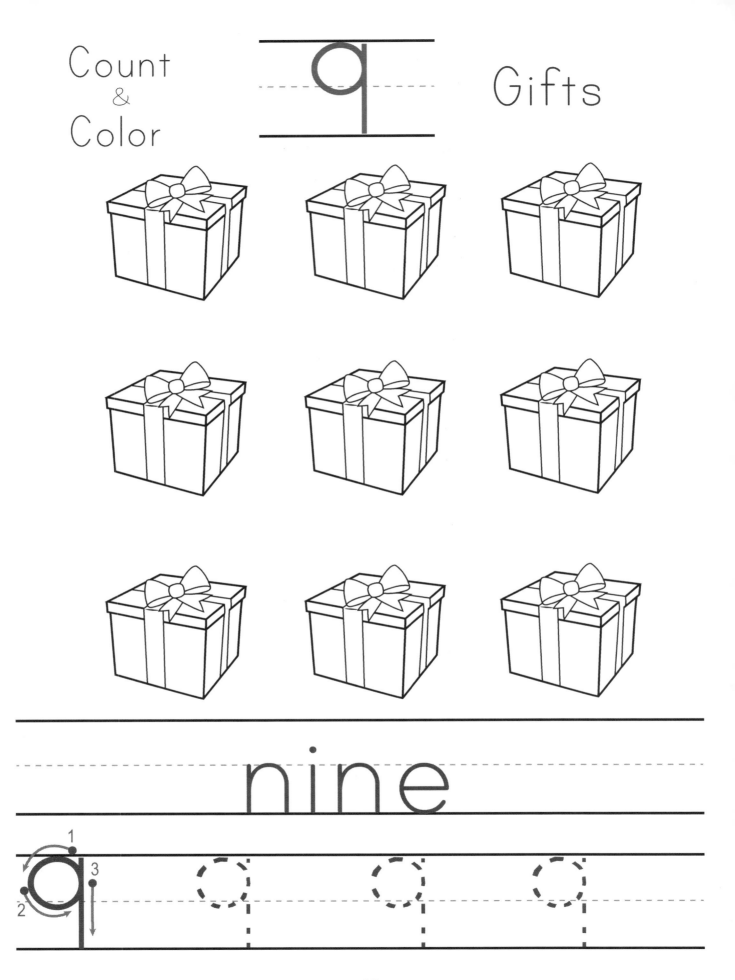

nine

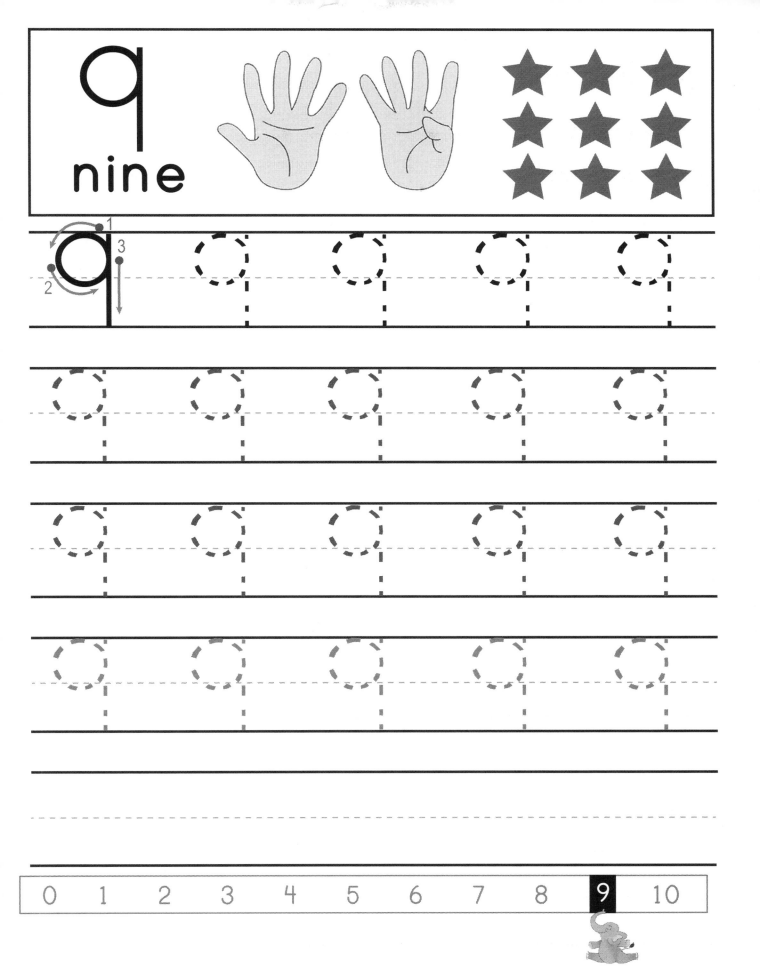

nine

0 1 2 3 4 5 6 7 8 **9** 10

Count
&
Color

10

Turtles

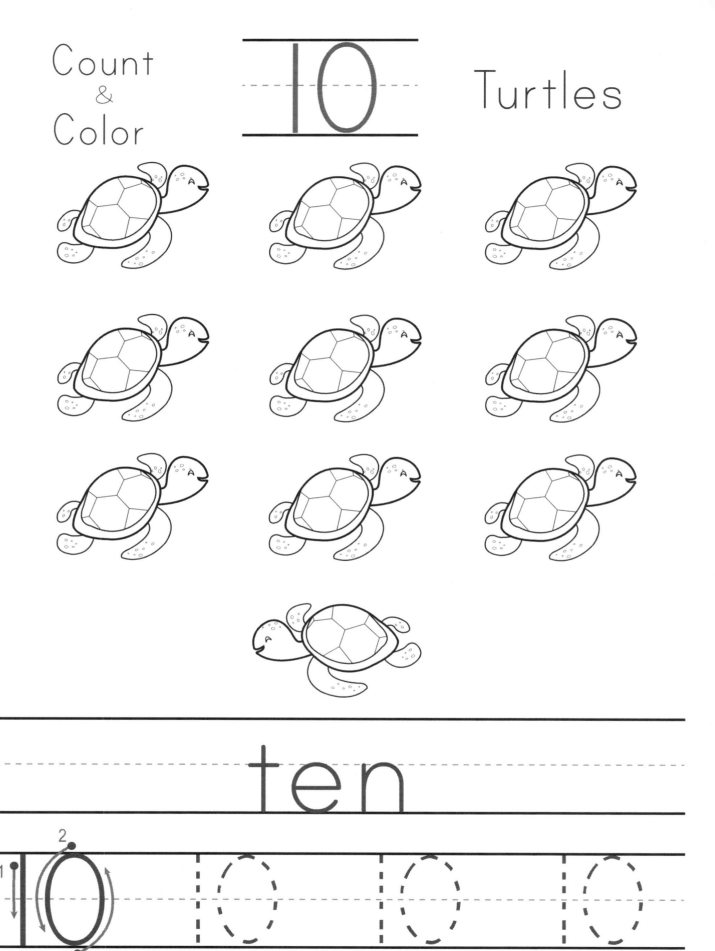

ten

24

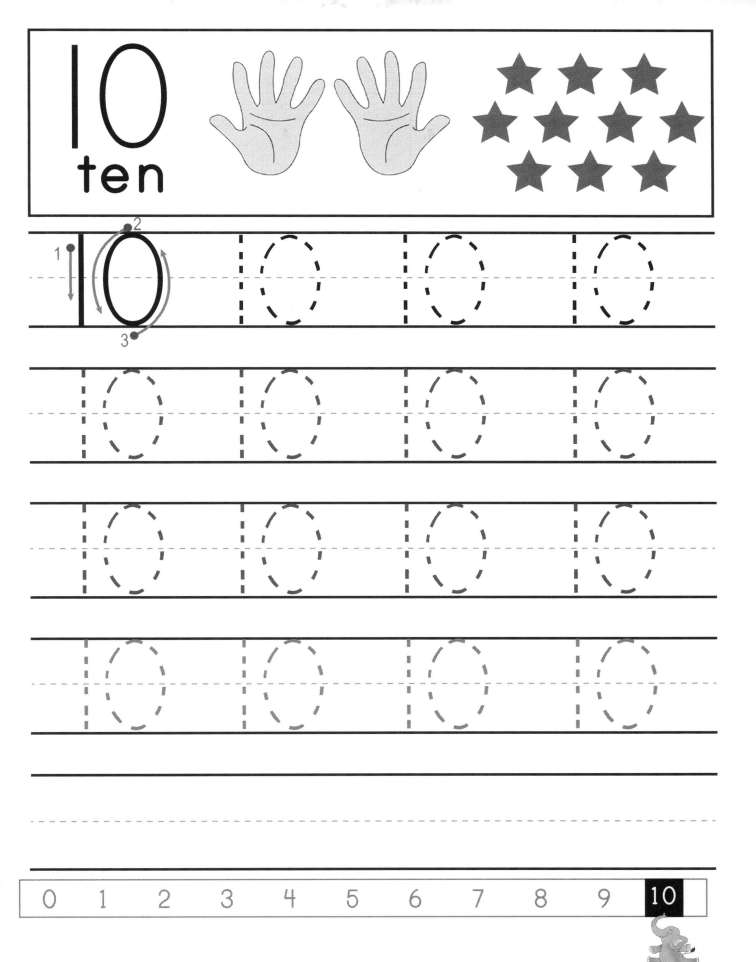

COLOR THE NUMBERS

Color the Circles with Number 0

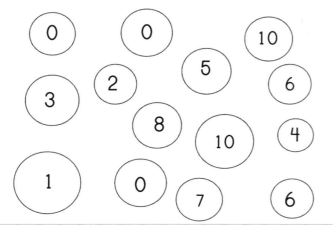

Color the Squares with Number 1

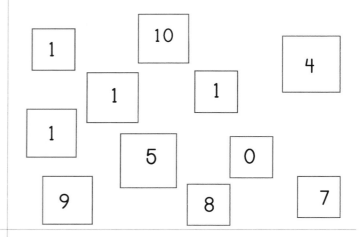

Color the Rectangles with Number 2

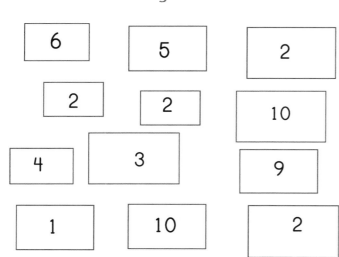

Color the Ovals with Number 3

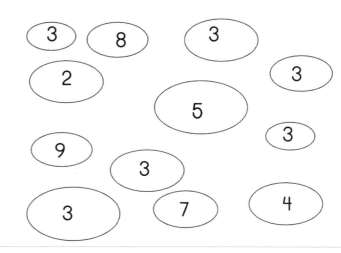

Color the Rhombus with Number 4

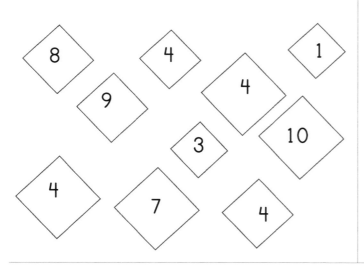

Color the Stars with Number 5

COLOR THE NUMBERS

Color the Hearts with Number 6

Color the Triangles with Number 7

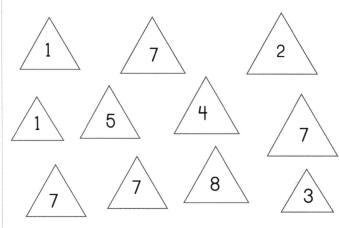

Color the Pentagon with Number 8

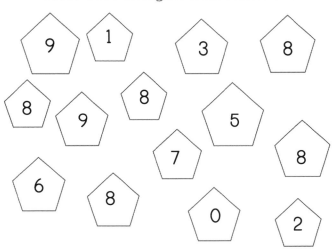

Color the Hexagon wtih Number 9

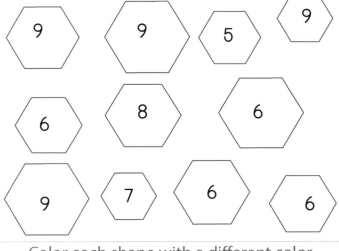

Color the Octogon with Number 10

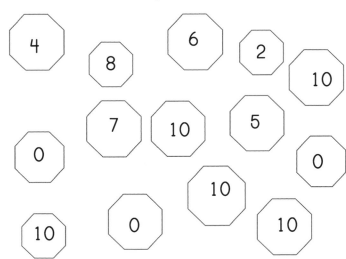

Color each shape with a different color

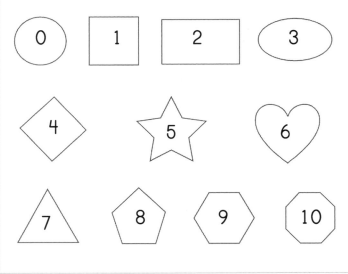

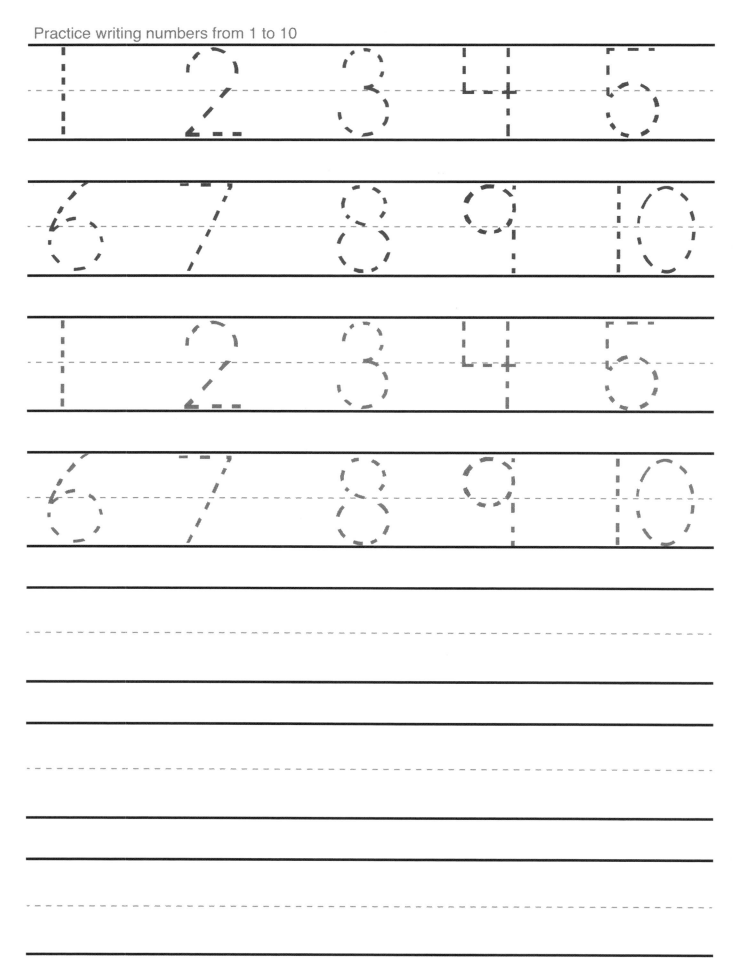

CONNECT THE DOTS

Draw the cat's tail by connecting the numbered dots from 1-10

COLOR BY NUMBER
Color the gift by using the Color Chart

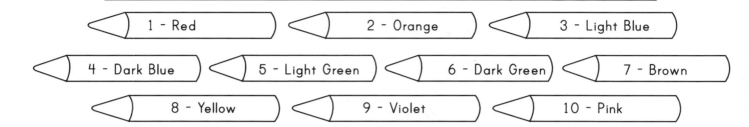

1 - Red　　2 - Orange　　3 - Light Blue

4 - Dark Blue　　5 - Light Green　　6 - Dark Green　　7 - Brown

8 - Yellow　　9 - Violet　　10 - Pink

Part 2:

Counting

Show numbers with your fingers.
Trace and practice the numbers.
Try writing them on your own on the blank line.

Shapes & Objects

Learn to trace shapes & count them.

Before, Between & After

Practice number sequence

You are AWESOME!

COUNTING
Count the objects & trace the number
Show the number using your fingers

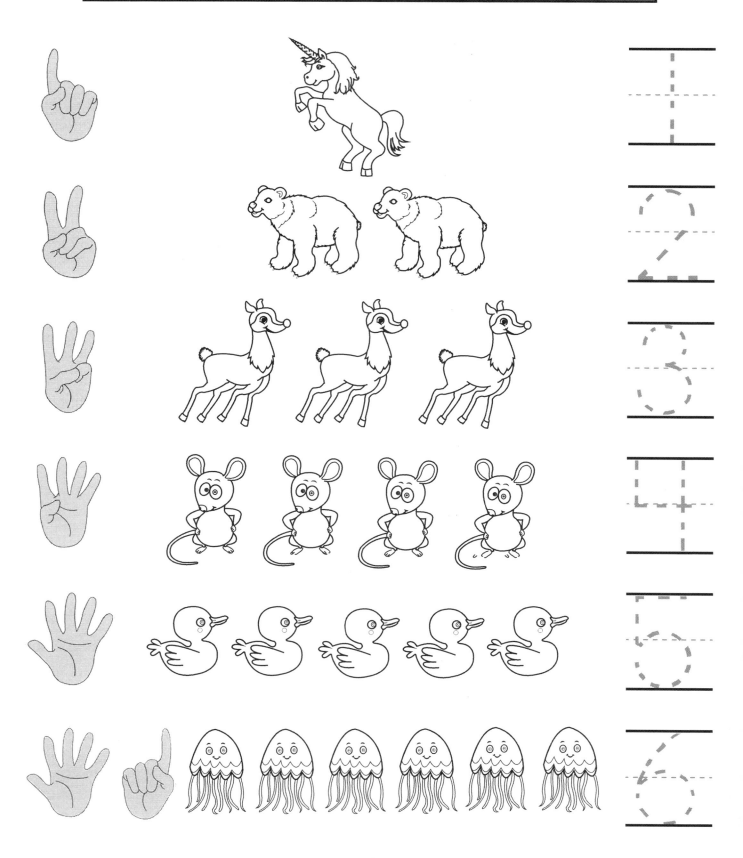

COUNTING
Count the objects & trace the number
Show the number using your fingers

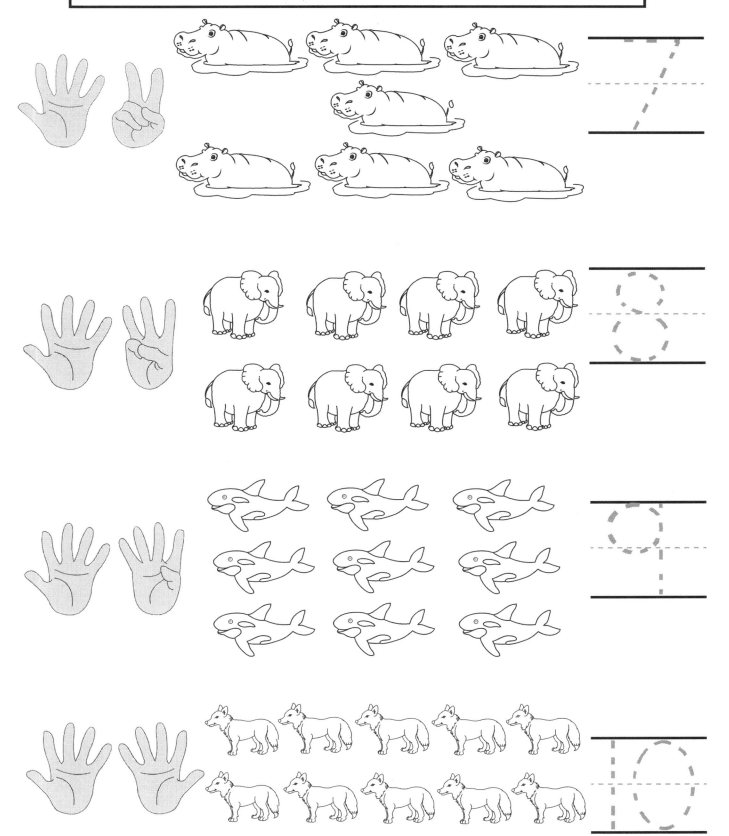

COUNTING DICE
Count the dots on the dice & trace the number
Count & color the stars

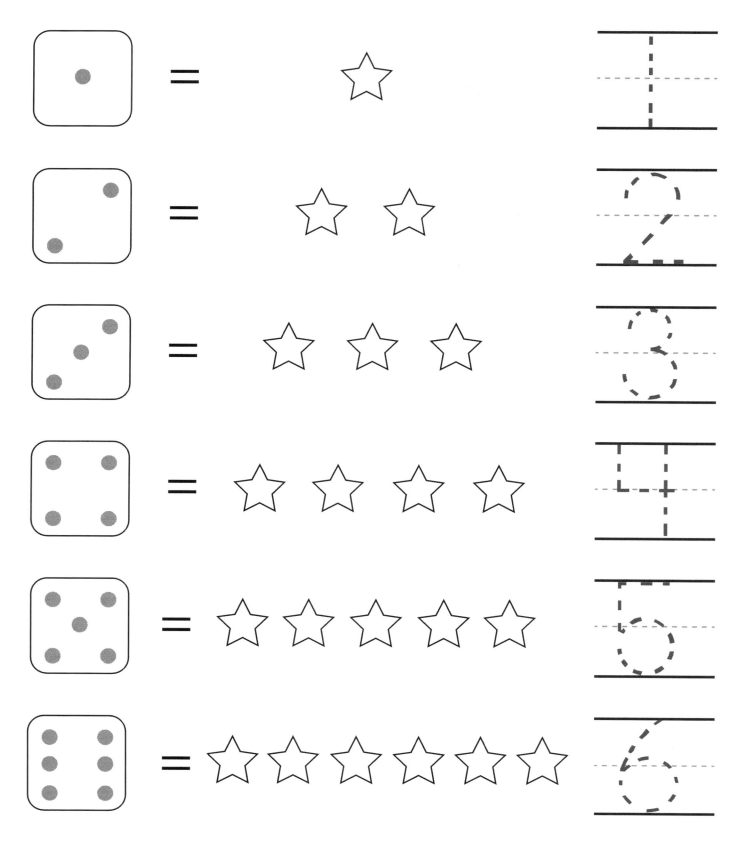

SHAPES

Count & trace the shapes

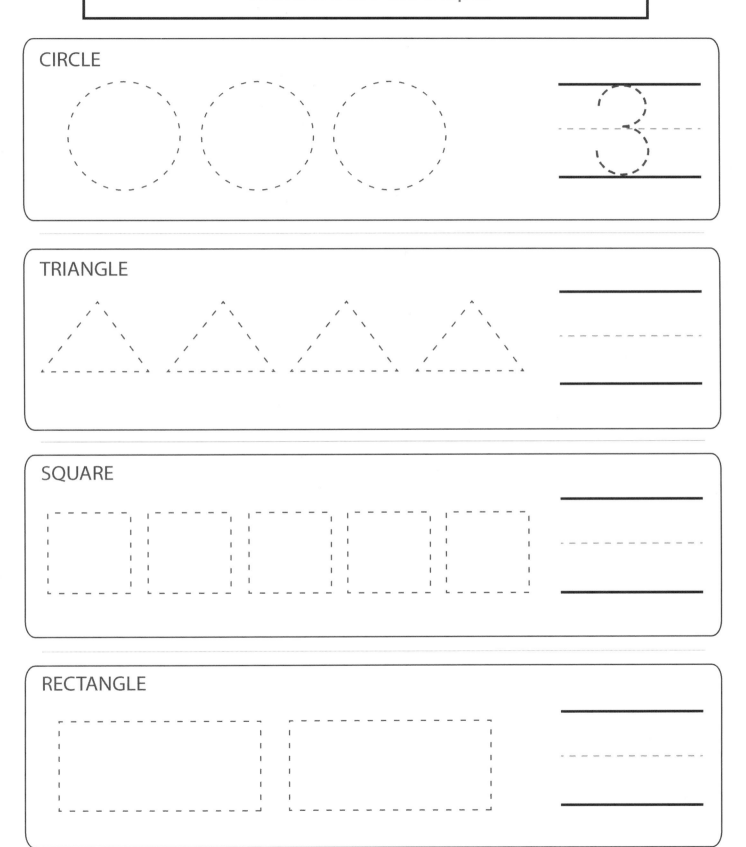

CIRCLE

TRIANGLE

SQUARE

RECTANGLE

SHAPES
Count & trace the shapes

RHOMBUS/DIAMOND

OVAL

HEART

STAR

Circle the number that matches the picture.
Color your favorite candies

1 2 ③ 4 5

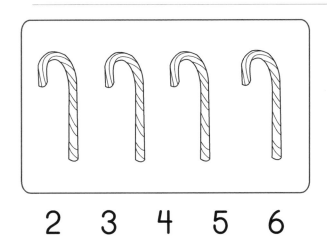

5 6 7 8 9

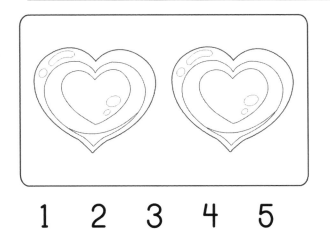

1 2 3 4 5

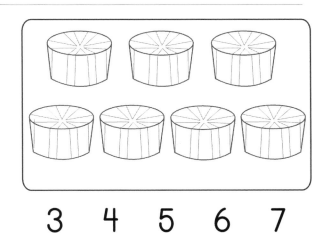

3 4 5 6 7

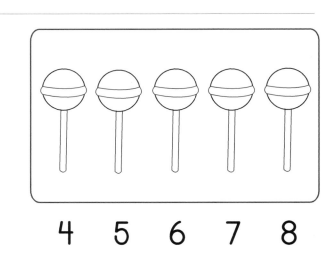

2 3 4 5 6

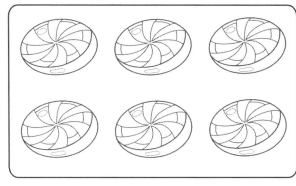

4 5 6 7 8

Circle the number that matches the picture.
Color your favorite flowers

5 6 7 8 9

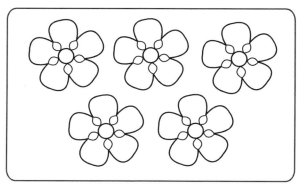

2 3 4 5 6

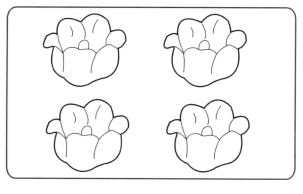

1 2 3 4 5

3 4 5 6 7

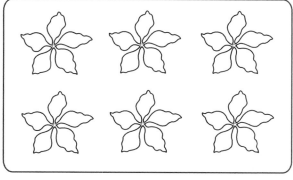

4 5 6 7 8

6 7 8 9 10

Count and circle the correct number in each row
Which ocean animal is your favorite?

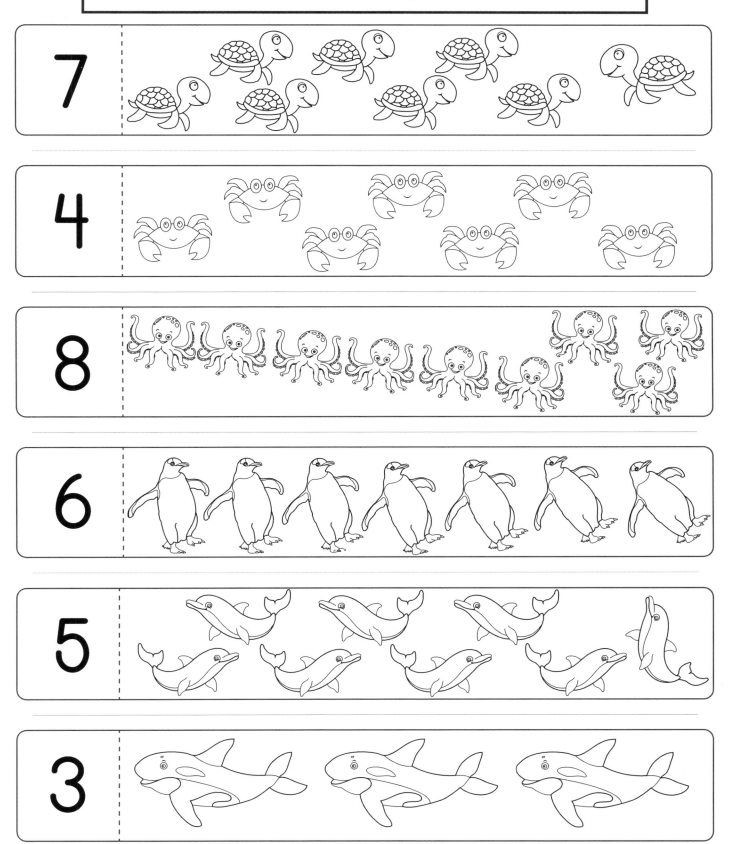

Count and circle the correct number in each row
Color your favorite magical animal

Count the shells in the sea.
Color the sea that has 4 shells.

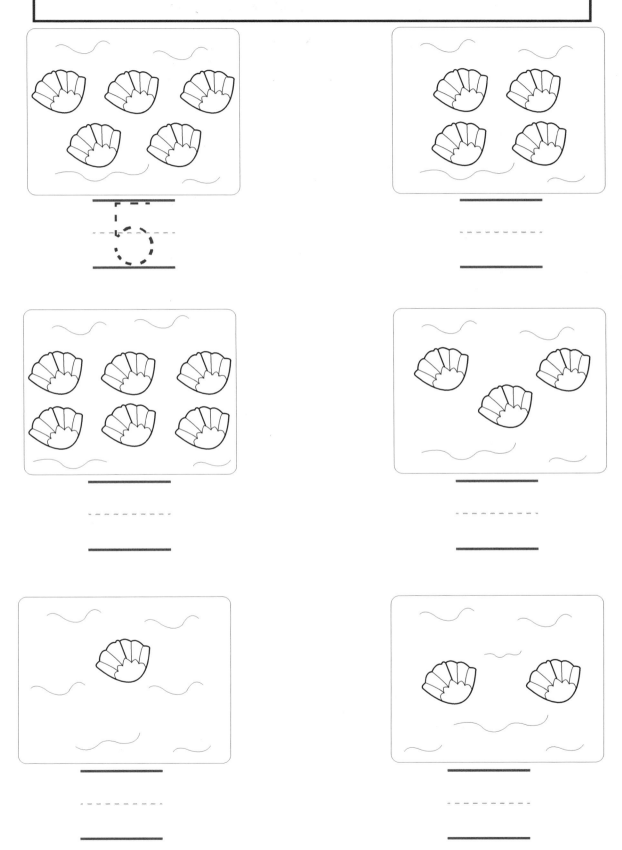

Count the apples on the tree.
Color the tree that has 0 apples.

Count the fish in each bowl.
Color the bowl that has 3 fish.

Count the flowers in each bush.
Color the bush that has 9 flowers.

Part 3:

Match objects

Learn to count number of objects

Before, Between & After

Practice number sequence

One More & One Less

Recognize which group has more/less

You are AWESOME!

SHAPES
Match the numbers & trace them

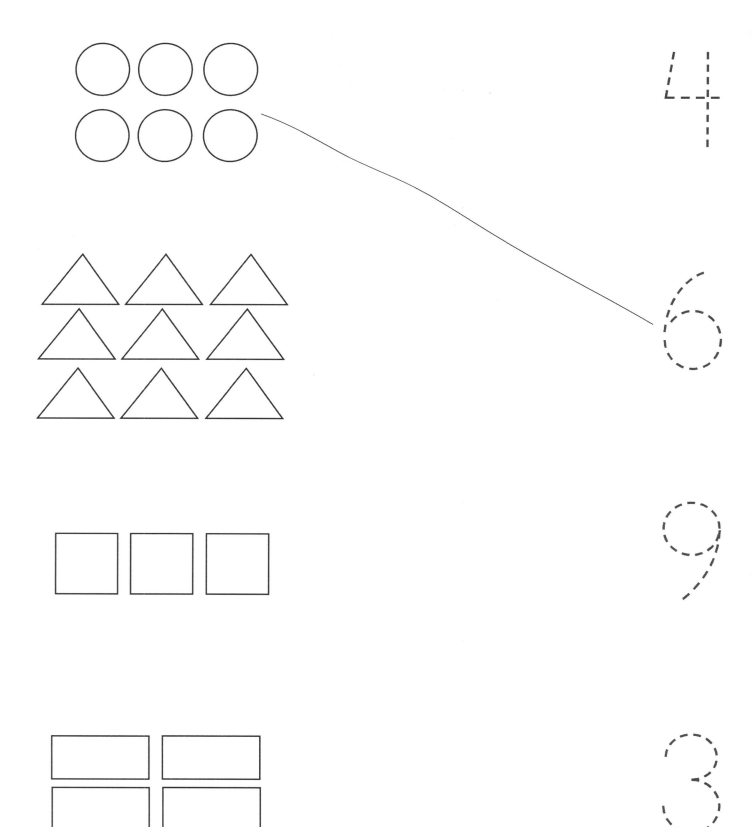

SHAPES
Match the numbers & trace them

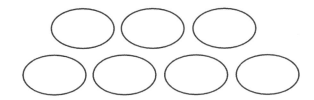 

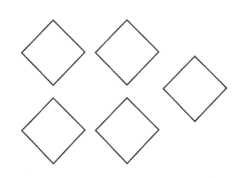 

ANIMALS
Match the numbers & trace them

FRUITS
Match the numbers & trace them

1   2   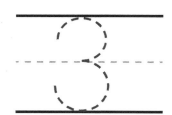

2   3   _____

3   4   _____

4   5   _____

AFTER

Write the number that comes **after** these numbers

5   6   _____

6   7   _____

7   8   _____

8   9   _____

BETWEEN

Write the number that comes **between** these numbers

1  3

2 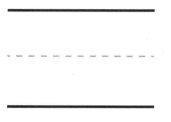 4

3  5

4 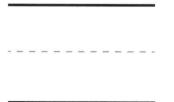 6

BETWEEN
Write the number that comes **between** these numbers

5 _____ 7

6 _____ 8

7 _____ 9

8 _____ 10

BEFORE

Write the number that comes **before** these numbers

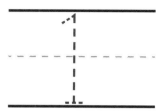 2   3

_____   3   4

_____   4   5

_____   5   6

BEFORE
Write the number that comes **before** these numbers

_____   6 7

_____   7 8

_____   8 9

_____   9 10

BEFORE, BETWEEN & AFTER
Fill in the blanks with the correct number

4   5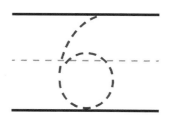

2   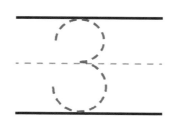   4

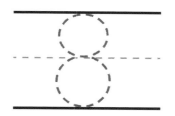   9   10

7   8   ____

BEFORE, BETWEEN & AFTER
Fill in the blanks with the correct number

3   4   ___

0   ___   2

___   8   9

4   ___   6

BEFORE, BETWEEN & AFTER

Fill in the blanks with the correct number

_____   5   6

3   _____   5

6   7   _____

_____   4   5

COUNT ONE MORE

Count the fish in each group.
Put a check (✔) in the box on the group that has one more

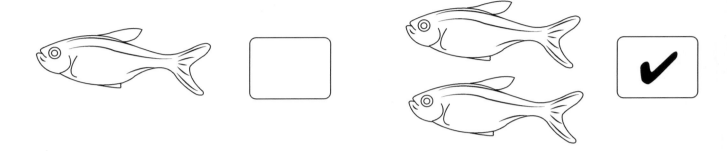

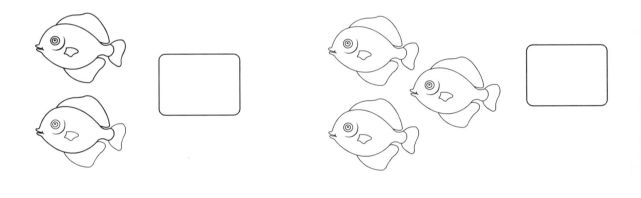

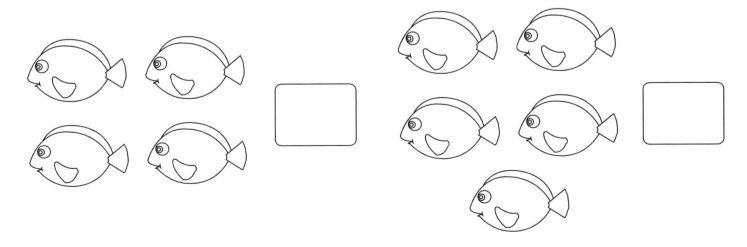

COUNT ONE MORE

Count the flowers in each group.
Put a check (✔) in the box on the group that has one more

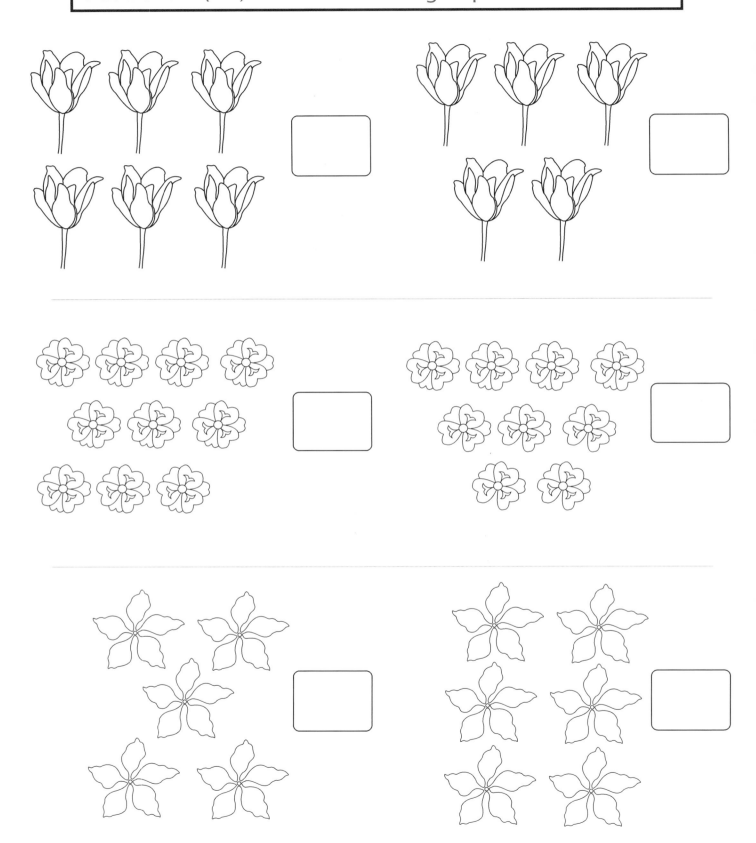

COUNT ONE LESS

Count the unicorns in each group.
Put a check (✔) in the box on the group that has one less

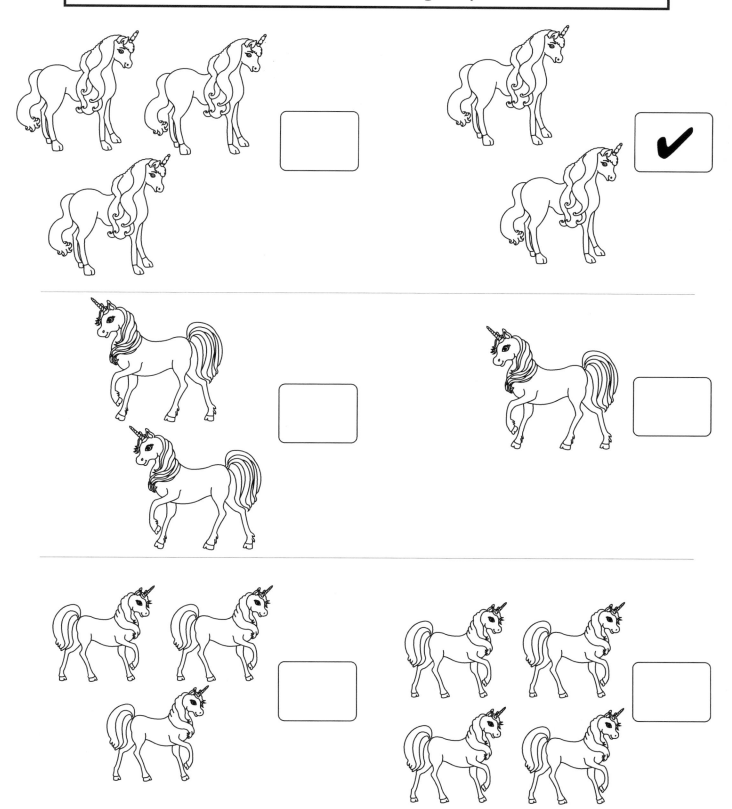

COUNT ONE LESS

Count the cupcakes in each group.
Put a check (✔) in the box on the group that has one less

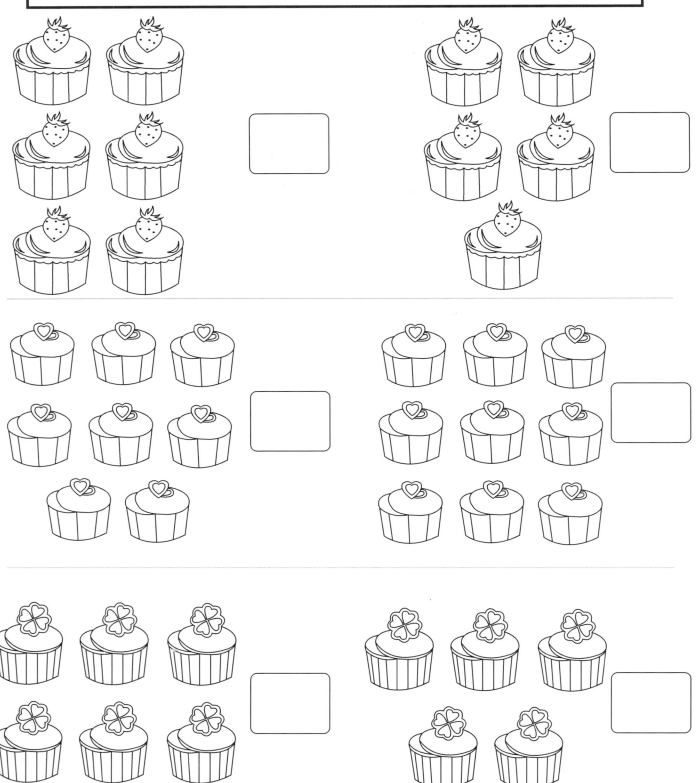

Part 4:

Directions & Manipulatives

Left & Right Differentiation

Follow Directions

Find Patterns

MATCHING DIRECTIONS
Practice recognizing Left & Right

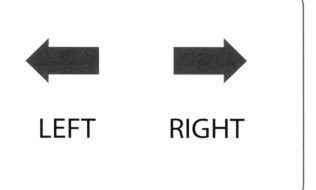

LEFT   **RIGHT**

Practice recognizing left & right
by looking at these arrows

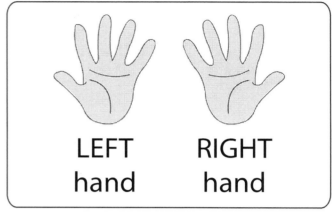

LEFT hand   **RIGHT hand**

Practice recognizing left & right
by looking at your palms

Color the duck on the left

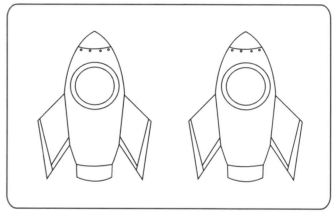

Color the rocket on the right

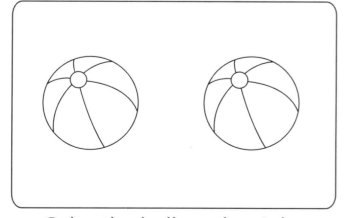

Color the ball on the right

Color the umbrella on the left

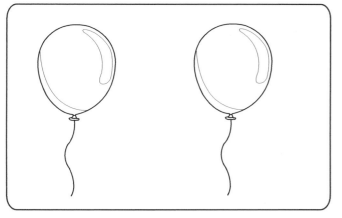

Color the balloon on the left

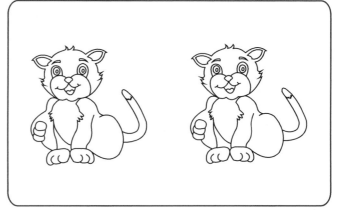

Color the cat on the right

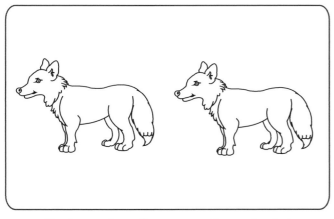

Color the fox on the right

Color the carrot on the left

Color the bow on the left

Color the fish on the right

MATCHING DIRECTIONS
Look at the first arrow on each line.
Draw a circle around the arrow that is looking the same way

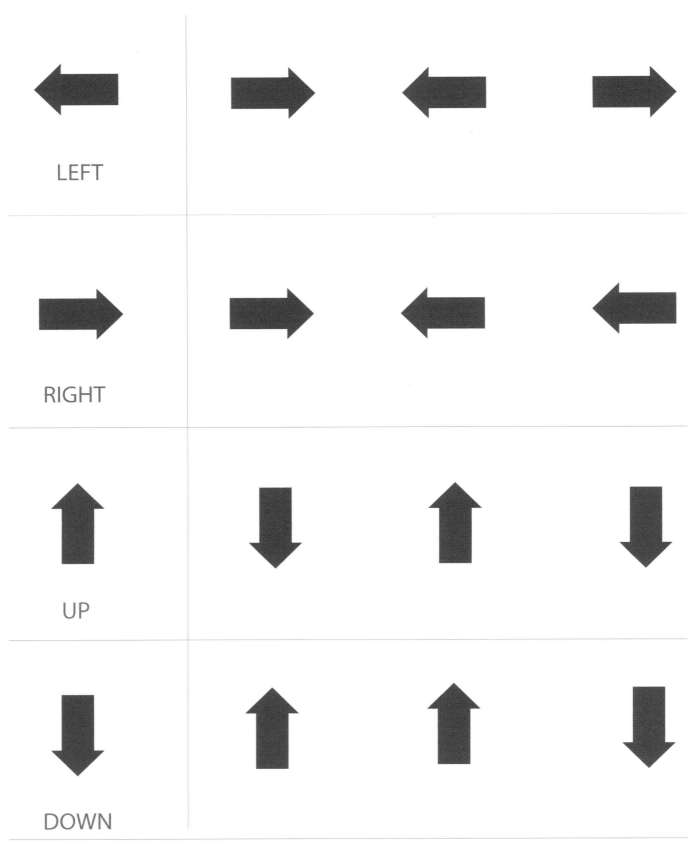

LEFT

RIGHT

UP

DOWN

MATCHING DIRECTIONS

Look at the first animal on each line.
Draw a circle around the animal that is looking the same way

MATCHING DIRECTIONS

Look at the first animal on each line.
Draw a circle around the animal that is looking the same way

PATTERN
Color the shape that comes next in the pattern

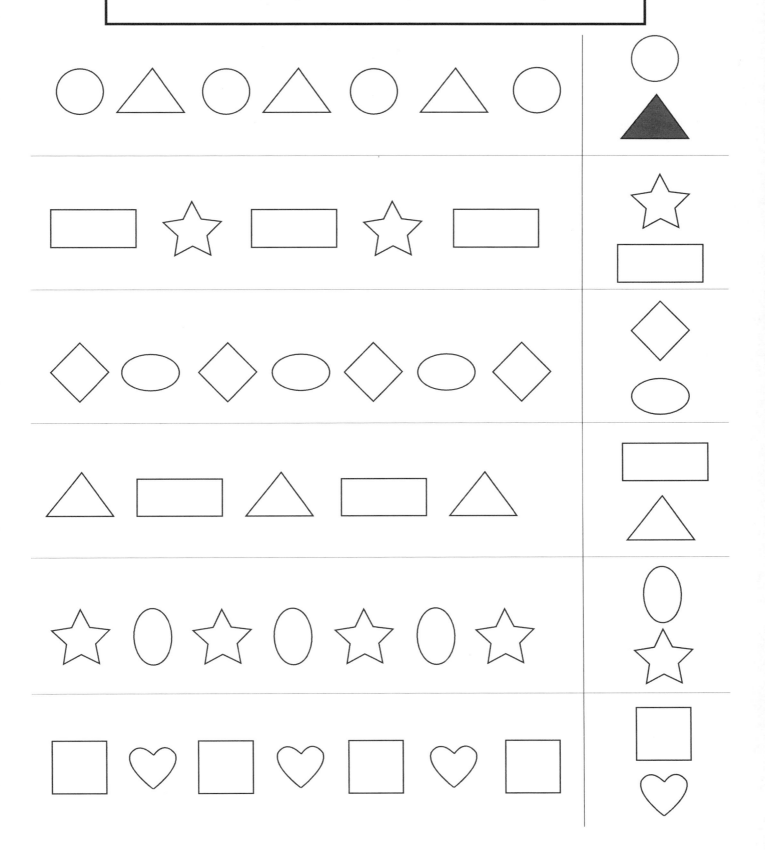

PATTERN
Color the shape that comes next in the pattern

PATTERN
Spot the pattern and fill in the missing patterns

Write the next two numbers in this pattern

5 6 5 6 5 6 5 6 |

Write the next three numbers in this pattern

1 2 3 1 2 3 1 |

Complete the next two letters in this pattern

Y Z Y Z Y Z Y Z |

Complete the the next three letters in this pattern

B C A B C A B |

Complete the next two dice patterns

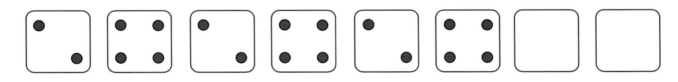

Complete the next two dice patterns

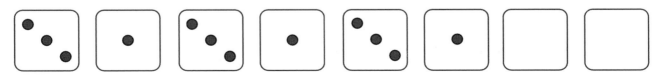

Part 5:

Simple Addition

Add the numbers.
Use the objects to count the total.
Trace and write down the answer.

You are
SENSATIONAL!

ADDITION
Count the stars
Write down the total number of stars

1 ⭐ + 1 ⭐ = 2 ⭐⭐

2 ⭐⭐ + 1 ⭐ = 3 ⭐⭐⭐

3 ⭐⭐⭐ + 1 ⭐ = 4 ⭐⭐⭐⭐

4 ⭐⭐⭐⭐ + 1 ⭐ = 5 ⭐⭐⭐⭐⭐

ADDITION
Count the stars
Write down the total number of stars

5   +   3   =   8

6   +   2   =

7   +   3   =

8   +   1   =

ADDITION
Count the circles
Write down the total number of circles

9 + 1 =

4 + 2 =

6 + 3 =

7 + 1 =

Count & Add the flowers

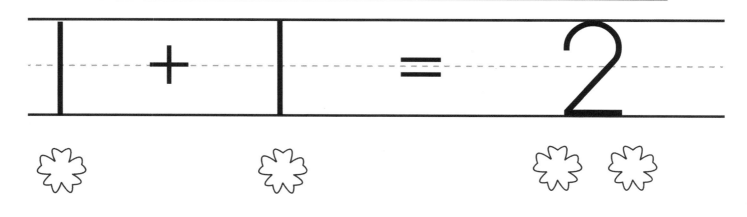

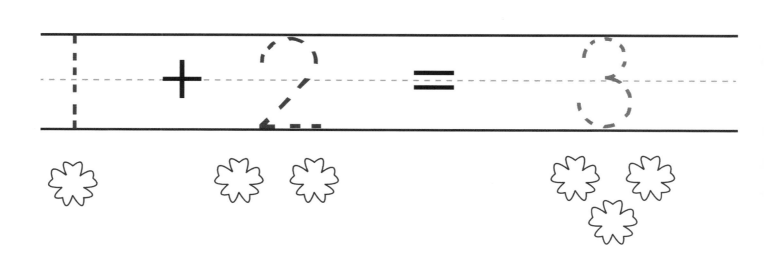

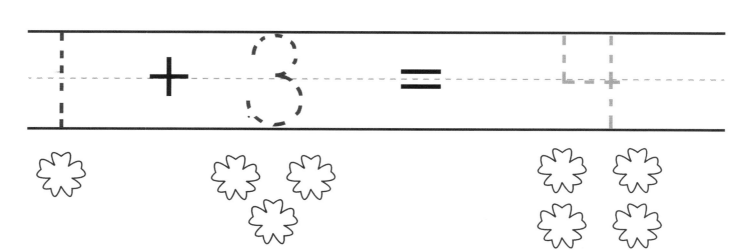

Count & Add the pizza slices

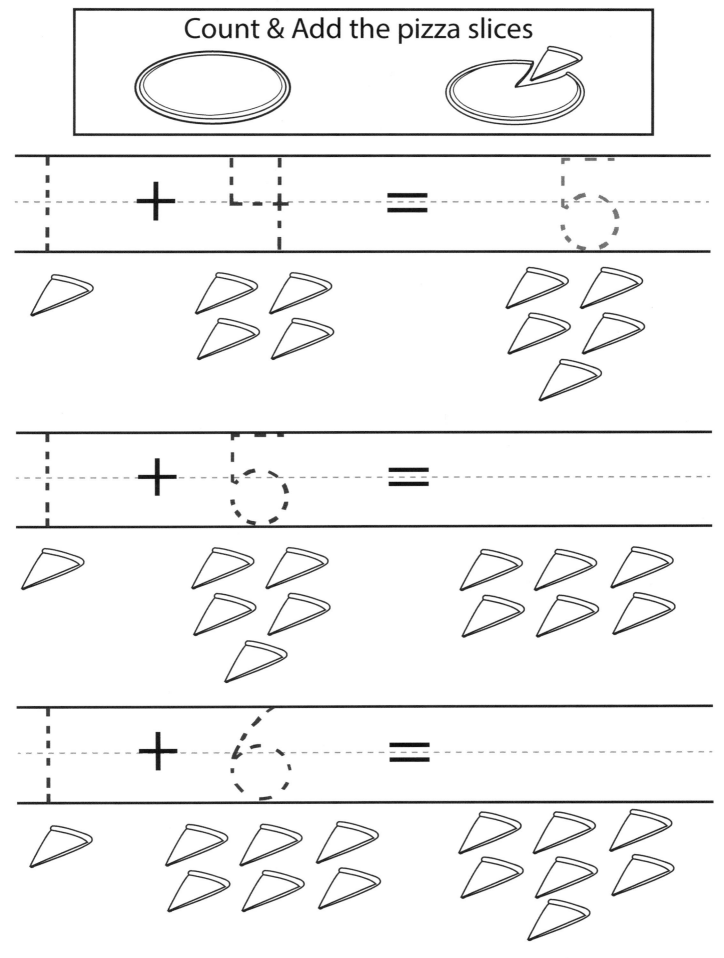

77

Count & Add the rockets

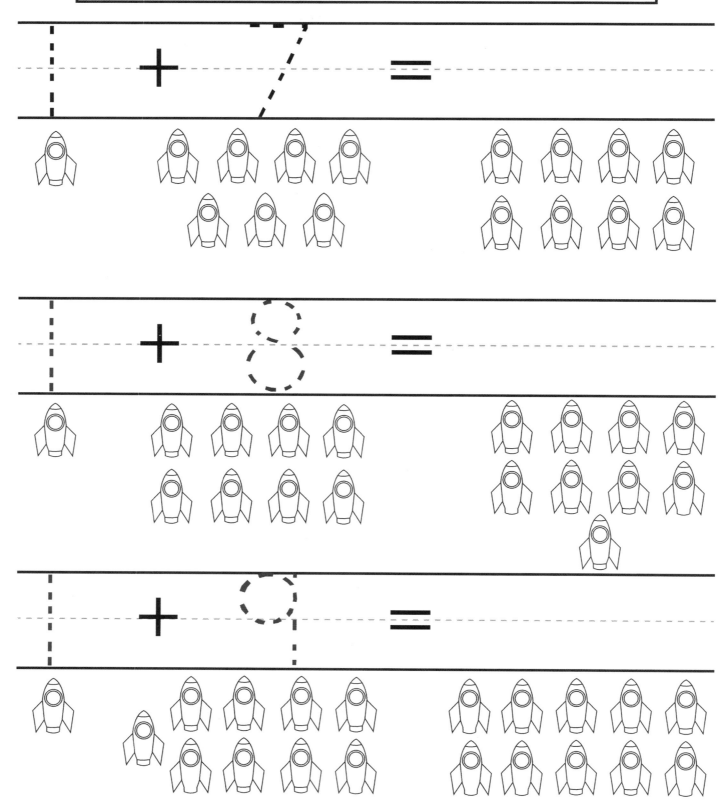

Count & Add the cookies

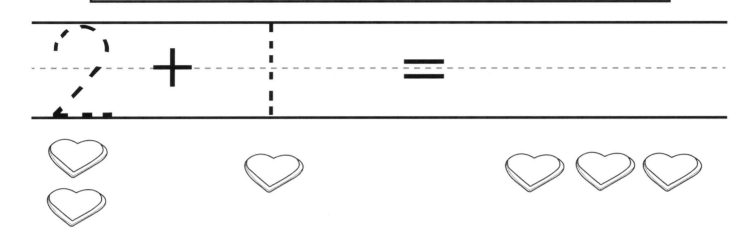

$$2 + 1 =$$

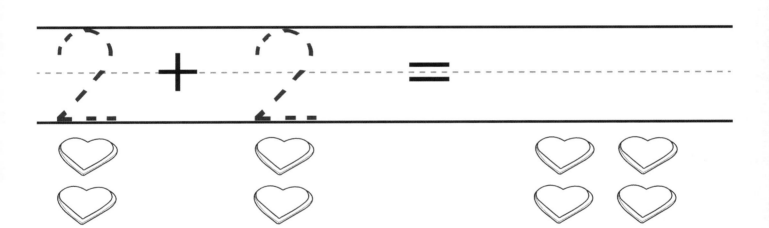

$$2 + 2 =$$

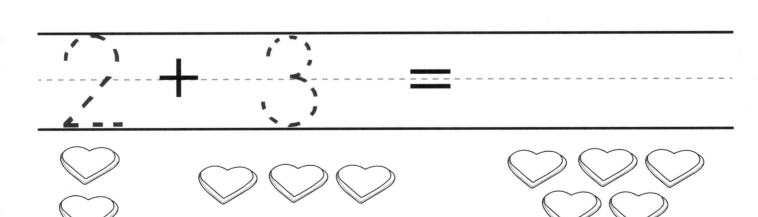

$$2 + 3 =$$

Count & Add the cupcakes

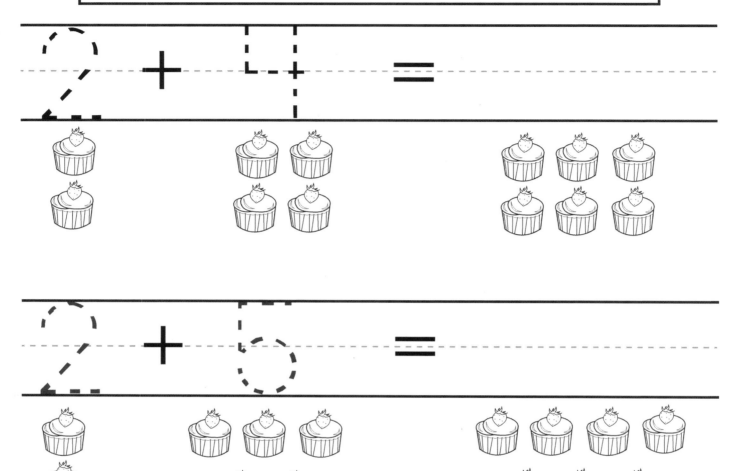

2 + [ ] =

2 + 6 =

2 + 6 =

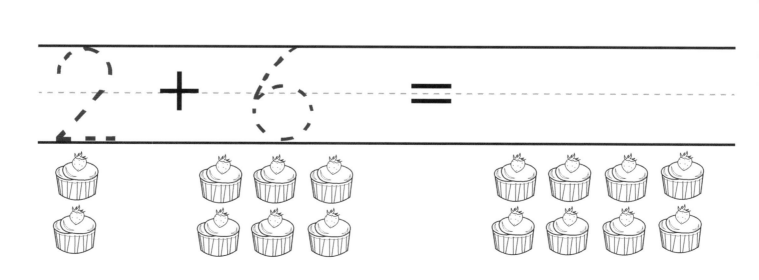

Count, add and write the total number of apples

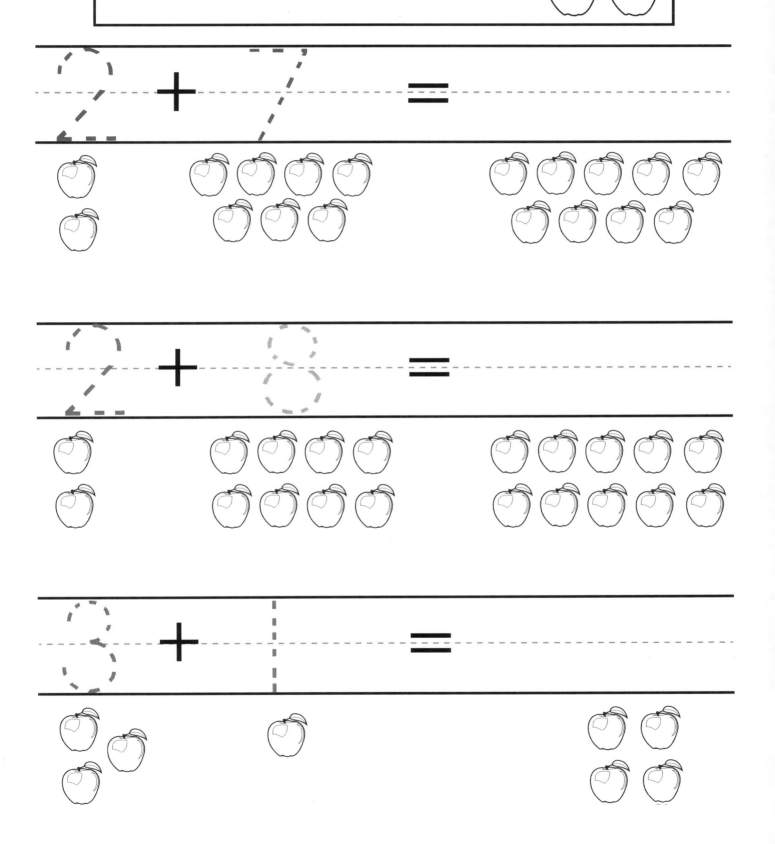

Count, add and write the total number of hats

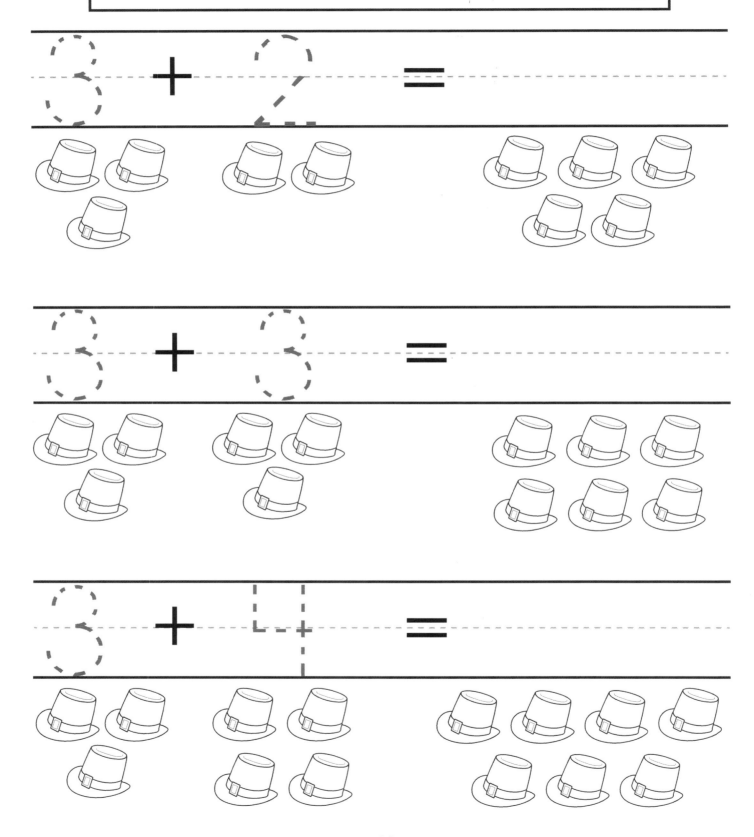

$3 + 2 =$

$3 + 3 =$

$3 + 4 =$

82

Count, add and write the total number of trees

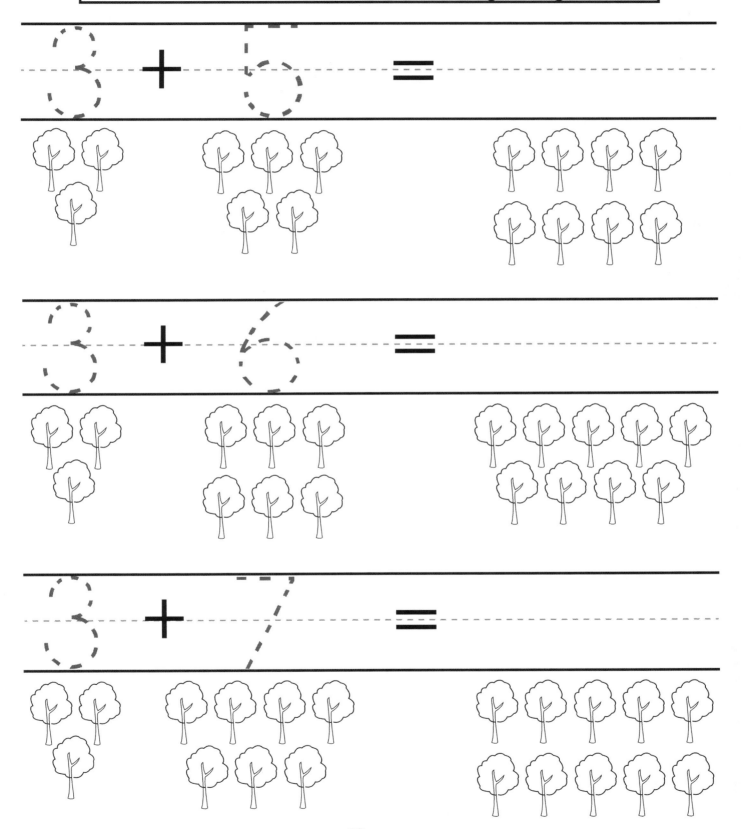

Count, add and write the total number of pears

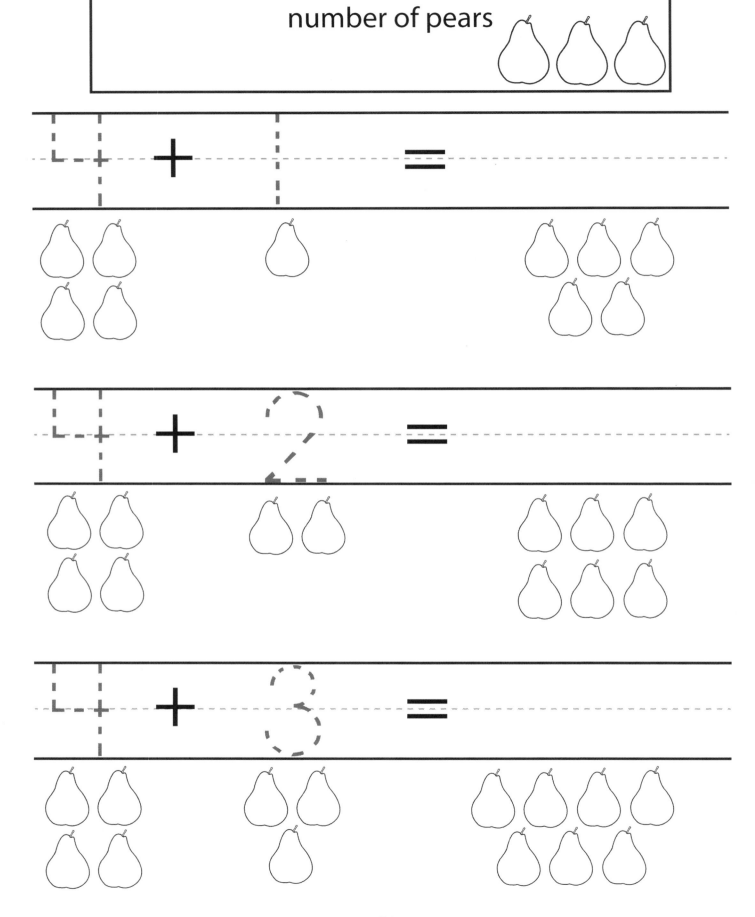

Count, add and write the total number of pencils

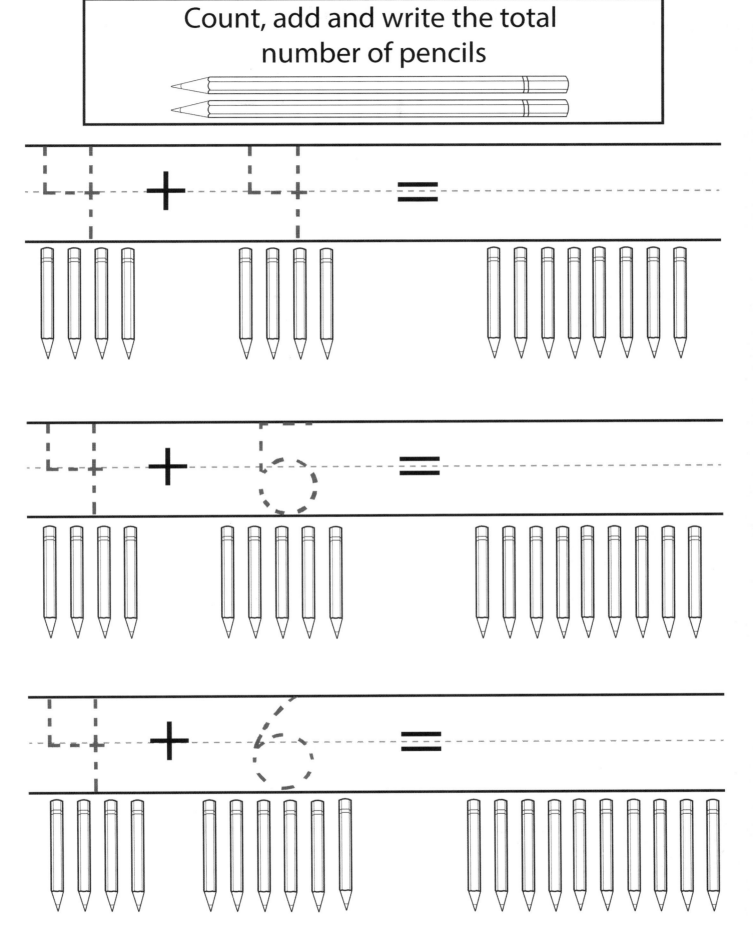

Count, add and write the total number of frogs

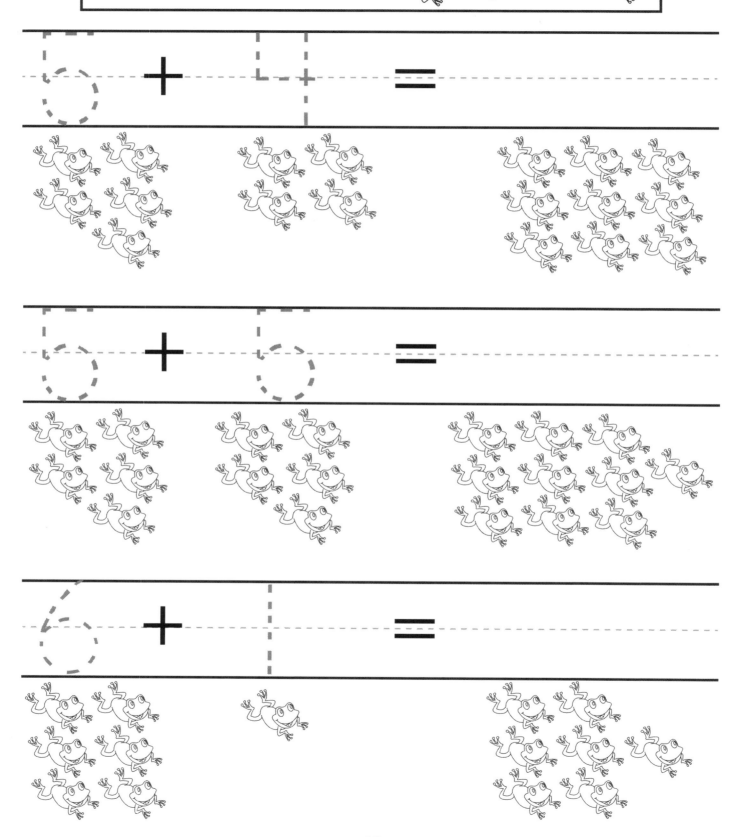

Count, add and write the total number of cherries

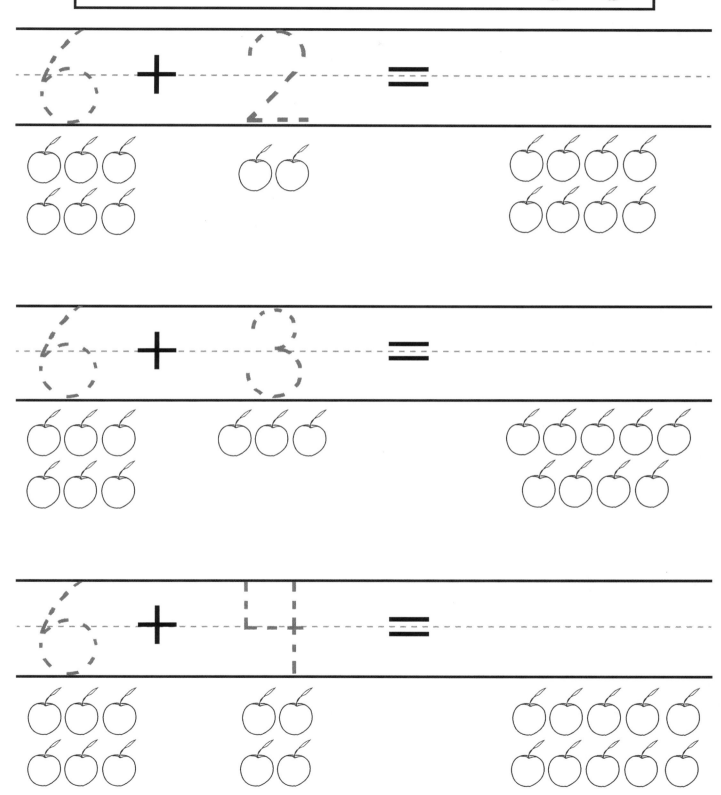

Count, add and write the total number of crayons

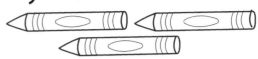

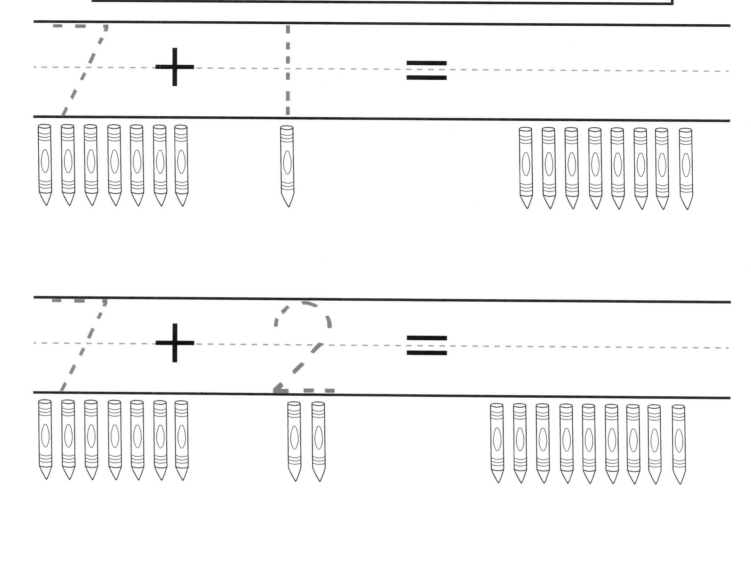

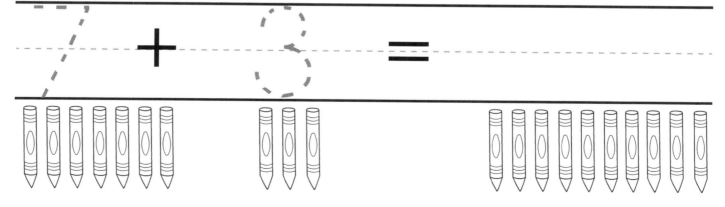

Part 6:

Simple Subtraction

Learn to subtract numbers using objects

SUBTRACTION
Count the stars, cross out the number of stars you need to subtract.
Write how many stars are left in the space given on the right

   $2 - 1 = \underline{1}$

⭐⭐⭐ $3 - 1 = \underline{2}$

⭐⭐⭐ $3 - 2 = \underline{1}$

⭐⭐⭐ $3 - 3 = \underline{0}$

SUBTRACTION

Count the stars, cross out the number of stars you need to subtract.
Write how many stars are left in the space given on the right

★ ★ ★ ★ $4-1=$ _3_

★ ★ ★ ★ $4-2=$ _2_

★ ★ ★ ★ $4-3=$ _1_

★ ★ ★ ★ $4-4=$ _0_

SUBTRACTION

Count the stars, cross out the number of stars you need to subtract.
Write how many stars are left in the space given on the right

★ ★ ★ ★ ★   5 - 1 = 4

★ ★ ★ ★ ★   5 - 2 = 3

★ ★ ★ ★ ★   5 - 3 = 2

★ ★ ★ ★ ★   5 - 4 = 1

SUBTRACTION

Count the stars, cross out the number of stars you need to subtract.
Write how many stars are left in the space given on the right

★ ★ ★ ★ ★ ★ $6 - 1 =$ _____

★ ★ ★ ★ ★ ★ $6 - 2 =$ _____

★ ★ ★ ★ ★ ★ $6 - 3 =$ _____

★ ★ ★ ★ ★ ★ $6 - 4 =$ _____

SUBTRACTION

Count the stars, cross out the number of stars you need to subtract.
Write how many stars are left in the space given on the right

★ ★ ★ ★ ★ ★ $6-5=$ _____

★ ★ ★ ★ ★ ★ $6-6=$ _____

★ ★ ★ ★
★ ★ ★ $7-1=$ _____

★ ★ ★ ★
★ ★ ★ $7-2=$ _____

SUBTRACTION

Count the stars, cross out the number of stars you need to subtract.
Write how many stars are left in the space given on the right

$7 - 3 =$ _____

$7 - 4 =$ _____

$7 - 5 =$ _____

$7 - 6 =$ _____

SUBTRACTION

Count the stars, cross out the number of stars you need to subtract.
Write how many stars are left in the space given on the right

$7 - 7 =$ _____

$8 - 1 =$ _____

$8 - 2 =$ _____

$8 - 3 =$ _____

SUBTRACTION

Count the stars, cross out the number of stars you need to subtract.
Write how many stars are left in the space given on the right

$8 - 4 =$ _____

$8 - 5 =$ _____

$8 - 6 =$ _____

$8 - 7 =$ _____

SUBTRACTION
Count the stars, cross out the number of stars you need to subtract.
Write how many stars are left in the space given on the right

$8 - 8 =$ _____

$9 - 1 =$ _____

$9 - 2 =$ _____

$9 - 3 =$ _____

SUBTRACTION

Count the stars, cross out the number of stars you need to subtract.
Write how many stars are left in the space given on the right

$9 - 4 =$ _____

$9 - 5 =$ _____

$9 - 6 =$ _____

$9 - 7 =$ _____

SUBTRACTION

Count the stars, cross out the number of stars you need to subtract.
Write how many stars are left in the space given on the right

$9 - 8 =$ _____

$9 - 9 =$ _____

$10 - 1 =$ _____

$10 - 2 =$ _____

SUBTRACTION

Count the stars, cross out the number of stars you need to subtract.
Write how many stars are left in the space given on the right

$10-3 = \underline{}$

$10-4 = \underline{}$

$10-5 = \underline{}$

$10-6 = \underline{}$

SUBTRACTION

Count the stars, cross out the number of stars you need to subtract.
Write how many stars are left in the space given on the right

$10 - 7 =$ ___

$10 - 8 =$ ___

$10 - 9 =$ ___

$10 - 10 =$ ___

SUBTRACTION
Count the unicorns, cross out the number of unicorns you need to subtract.
Write how many you have left in the space given on the right

$$5 - 3 = \underline{}$$

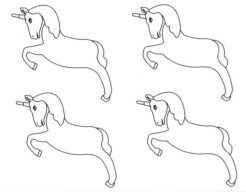

$$4 - 3 = \underline{}$$

$$3 - 0 = \underline{}$$

$$2 - 2 = \underline{}$$

SUBTRACTION

Count the candies, cross out the number of candies you need to subtract.
Write how many you have left in the space given on the right

4 - 2 = _____

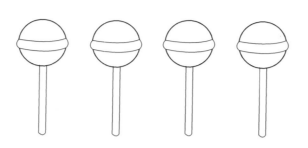

3 - 1 = _____

4 - 1 = _____

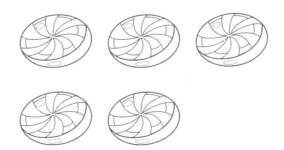

5 - 2 = _____

SUBTRACTION

Count the fruits, cross out the number of fruits you need to subtract.
Write how many you have left in the space given on the right

$$7 - 3 = \underline{\qquad}$$

$$9 - 4 = \underline{\qquad}$$

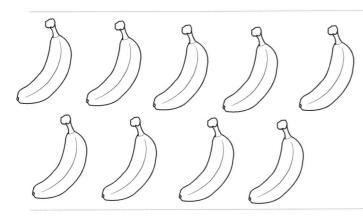

$$6 - 3 = \underline{\qquad}$$

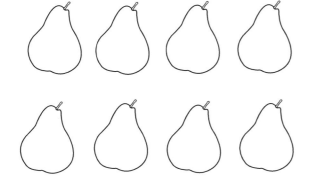

$$8 - 5 = \underline{\qquad}$$

SUBTRACTION

Count the flowers, cross out the number of flowers you need to subtract.
Write how many you have left in the space given on the right

$9 - 7 =$ _____

$8 - 2 =$ _____

$6 - 1 =$ _____

$7 - 5 =$ _____

SUBTRACTION

Count the leaves, cross out the number of leaves you need to subtract.
Write how many you have left in the space given on the right

$$7 - 6 = \underline{\quad}$$

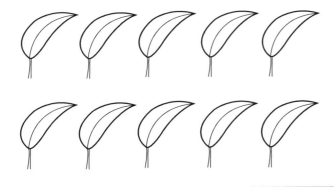

$$5 - 1 = \underline{\quad}$$

$$10 - 3 = \underline{\quad}$$

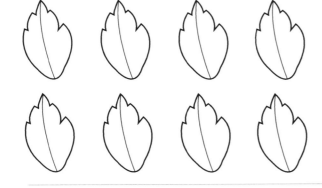

$$8 - 4 = \underline{\quad}$$

Preschool Math
Completed ✓

CONGRATULATIONS!
You are
AWESOME!

Recommended next skill

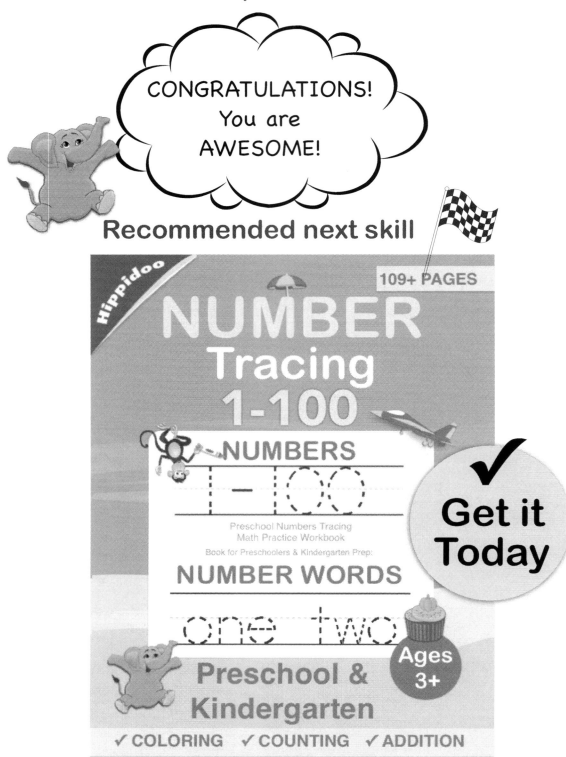
ISBN: 1692814761

Celebrate your Success!

Share the Joy!

Feel Great Everyday!

Write to me at **sujatha.lalgudi@gmail.com** with the subject:
Preschool Math along with **your kid's name** to receive:

- Additional practice worksheets.
- A name tracing worksheet so your kid can practice writing their own name.
- An Award Certificate in Color to gift your child!

Congratulations
Math Super Star
Awarded to

_____

For _____

Date _____   Signed _____

Made in the USA
Columbia, SC
17 September 2021